T0092081

The Well-Connected Animal

The Well-Connected Animal

Social Networks and the Wondrous Complexity of Animal Societies

Lee Alan Dugatkin

The University of Chicago Press
Chicago and London

The University of Chicago Press, Chicago 60637
The University of Chicago Press, Ltd., London

Published 2024
Printed in the United States of America

33 32 31 30 29 28 27 26 25 24 1 2 3 4 5

ISBN-13: 978-0-226-81878-8 (cloth)
ISBN-13: 978-0-226-81879-5 (e-book)
DOI: https://doi.org/10.7208/chicago/9780226818795.001.0001

Library of Congress Cataloging-in-Publication Data

Names: Dugatkin, Lee Alan, 1962– author.
Title: The well-connected animal : social networks and the wondrous complexity of animal societies / Lee Alan Dugatkin.
Description: Chicago : The University of Chicago Press, 2024. | Includes bibliographical references and index.
Identifiers: LCCN 2023029710 | ISBN 9780226818788 (cloth) | ISBN 9780226818795 (ebook)
Subjects: LCSH: Animal behavior. | Social behavior in animals. | Social networks.
Classification: LCC QL775 .D846 2024 | DDC 591.5—dc23/eng/20230713
LC record available at https://lccn.loc.gov/2023029710

⊗ This paper meets the requirements of ANSI/NISO Z39.48-1992 (Permanence of Paper).

For Aaron and Meena

Contents

Preface · ix

1. The Networked Animal · 1
2. The Ties That Bind · 19
3. The Food Network · 39
4. The Reproduction Network · 61
5. The Power Network · 75
6. The Safety Network · 93
7. The Travel Network · 109
8. The Communication Network · 123
9. The Culture Network · 139
10. The Health Network · 163

Afterword · 183

Acknowledgments · 187

Notes · 189

Index · 209

A gallery of photos follows page 82.

Preface

This is not a book about Twitter, Facebook, or any other social media platform. It won't give you tips for building your own network of friends or your business. Instead, I want to introduce you to other networks all around us: networks that permeate the natural world. This is a book about the networks formed between nonhuman animals. These networks stand as a tribute to the complexity, depth, and wonder of life in animal societies.

My path to social networks in animals began when I was in graduate school in the late 1980s. My dissertation work focused on testing new models of cooperative behavior, and what I found, in a nutshell, was that guppies paid attention to the behavior of other guppies and tempered their antipredator behavior based on how fish near them behaved.

To know my study species inside and out, I must have had read every animal behavior paper ever written (in English) on guppies by the end of my PhD. Many of those papers were on how females select their mates: generally speaking, female guppies prefer males that are colorful and relatively free of parasites. An assumption—sometimes tested, sometimes not—was that female mating preferences are based on genetic predispositions, which they most certainly are. But it struck me that there might be more to it than that. After all, if guppies paid attention to each other when engaging in antipredator behavior, maybe females were watching the mate choice of other females and using that information when selecting their own mates. A series of mate-choice experiments showed that they were.[1]

As with antipredator behavior, when selecting a mate, information on how others behaved mattered. A lot. Natural selection seems to have favored the ability to obtain, process, and act upon information about others in complex social venues where such information affects success or failure at some task. A few years later my PhD student Ryan Earley and I found just that in yet another behavioral venue: male green swordtail fish spied on the fights of other males and took note of whether a possible future opponent had demonstrated itself to be dangerous (or not).[2]

Almost all of this work was based on pairwise or three-way behavioral interactions. But, along with other animal behaviorists, I was starting to think that the behavior of animals in groups might—at least in principle—be affected, directly and indirectly, by *many* others, and so interactions among a pair or trio might be the tip of the iceberg when it comes to the intricate set of relationships among all members of a group. The trouble is that once you get beyond behavioral dynamics in pairs and trios, it gets very complicated, very quickly, and so, what with the other projects I had going on at the time, I didn't dig any deeper. Fortunately, others did. For the last twenty years or so, animal behaviorists—also called ethologists—all over the planet have been studying *social networks* in nonhumans. In these social networks, information flows, directly and indirectly, between network members, and how individuals behave can have consequences that ripple through a group.

Early on, there was skepticism in the animal behavior research community about whether social networks mattered.

The general consensus was that nonhumans display a rich array of complex social behaviors, but not *that* complex: not social network complex. That response is in line with what has happened many times before. We assume complex behavior must be uniquely human, and so little research explores the phenomenon in animals. But eventually someone stumbles upon an example. And once that example makes its way into the literature, people begin to realize what Darwin and others had long recognized: that many complex behaviors are found across the animal kingdom. Soon researchers are proposing that we redefine the behavior in question, so that, at least in principle, its occurrence is possible in nonhumans. Social networks should not be defined in terms of Facebook and Twitter and the like, but rather by whether information flows, directly and indirectly, between members of a group, and whether or not individuals act upon that information.

When researchers began to publish examples of social networks in animals in the early 2000s, the skepticism waned. Since then, animal behavior researchers have been building models and testing hypotheses—most often in the field—about *how* social networks operate, *why* they work, who gets what, who matters most and who not so much, and more. We've discovered that some members of an animal social network have prominent roles, acting as information hubs; others less so. Networks can be large or small. Some networks are primarily female, others primarily male, but most networks contain members of both sexes. Networks can be structured in part based on kinship, but often they are not.

Within the same population of animals, the social network associated with one behavior (e.g., foraging) might be similar to the social network underlying another behavior (e.g., mating), but more often it isn't, which is to say that there are usually layered, connected social networks in play simultaneously, with individuals playing different roles in each. What social network *analysis* does is to allow researchers to picture and understand social networks in nonhumans and to generate testable hypotheses about the dynamics of such networks. I'll explore all this, offering a view of the intriguing, often complex, and subtle dynamics driving animal social networks.

Researchers have come to discover that being embedded in social networks plays a critical role in almost every aspect of animal life: what they eat, how they protect themselves, whom they mate with, the dynamics of parent-offspring relations, power struggles, navigation, communication, play, cooperation, culture, and more. Microbes—some good, some not so good—also hitchhike rides on the animals in these social networks.

Work on social networks in animals is now so cutting edge that it's time to tell its story, along with the story of the scientists behind it all. Drawing on research in animal behavior, evolution, computer science, psychology, anthropology, genetics, and neurobiology, *The Well-Connected Animal* will explore social networks in giraffes, elephants, kangaroos, many primate and bird species, Tasmanian devils, whales, bats, field crickets, manta rays, and more. In each case, I will examine how and why scientists tested their hypotheses. In so doing, we will travel around the planet with researchers studying social networks in the wild everywhere from Australia to Asia, Africa to Europe, and North and Central America to South America.

I've talked at length with all the researchers introduced in the book, and I made a concerted effort to see that the work of women scientists researching animal social networks receives the attention it so deserves. Close to 45% of the studies covered were led (or co-led) by women. I also cover the work of younger researchers, which was not difficult given that so many social network studies are run by a new generation of animal behaviorists trained (often self-trained) in this method.

Above and beyond learning about the science of animal social networks per se, we will also visit the study sites where the science is being done for a behind-the-scenes peek at what goes on when researchers study social networks in the wild. We will get a feel for what it is like to wake up in Uganda and roll out of bed right into the midst of a chimpanzee social network, what happens when you try to track social networks in birds that are migrating thousands of kilometers, why on earth anyone would look for social networks in manta rays of all things, and much more. Along with the adventure of it all, we'll learn how serendipity often outflanks the best-laid plans, exploring the ups and downs and twists and turns of studying social animal networks in the wild.

1. The Networked Animal

In nature we never see anything isolated, but everything in connection with something else which is before it, beside it, under it and over it. —Johann Wolfgang von Goethe[1]

On the island of Cayo Santiago, about 1.5 kilometers off the coast of eastern Puerto Rico, the typical relationship between humans and other primates gets turned on its head. The 1,700 rhesus macaque monkeys (*Macaca mulatta*) living on that island have free rein to move around wherever and whenever they please. Humans don't. Following a fifteen-minute boat ride from a dock in the former fishing village of Punta Santiago, humans—two dozen of them at most—are allowed on the island starting at 7:30 a.m., at which point *they* step into cages for a few minutes, in part to drop off any food they have with them, before they are allowed out on the island. At human feeding time, around noon, they step back into the cages for lunch. Monkeys, on the other hand, are free to eat anywhere they please, including the food stations that are filled each morning by their human visitors. Starting about 2:30 p.m., all *Homo sapiens* become hominid non grata and are required to head to the dock, get back on the boat, and sail to the mainland, until they are allowed to return the next morning.

The Cayo macaque population came to be in 1938 when primatologist and explorer Clarence Carpenter captured about 500 rhesus macaques from twelve districts in India. He and the monkeys sailed from Calcutta via Boston and New York to San Juan, and from there to Cayo Santiago, a 15-hectare island that had recently been leased to the School of Tropical Medicine at the University of Puerto Rico. The *New York Times* regaled its readers with the story of the 51-day, 22,000-kilometer journey, during which Carpenter single-handedly looked after his macaque travel mates. *Life* magazine even saw fit to send Hansel Mieth, one of its most intrepid photographers, to Cayo to do a photo layout of the release of the macaques onto the island.[2]

Lauren Brent began working with macaques on Cayo Santiago when her PhD adviser suggested she consider going down to the island to have a look at the social dynamics of the monkeys. She immediately fell in love with everything about her study subjects, the island, and the questions she was soon addressing. "Cayo is a pretty perfect study system," Brent says, "and the macaques are such highly social little dudes." Her adviser pointed her to Darren Croft, Richard James, and Jens Krause's *Exploring Animal Social Networks*, the first book devoted to the subject in nonhumans, and soon Brent was thinking she could better understand her macaques using network thinking.[3]

When she finished her PhD work, Brent joined the lab of Michael Platt, a neuroscientist interested in primate brains, who had received a research grant from the National Institutes of Health to study the macaques on Cayo Santiago. He knew Brent had experience with the macaques there, and so he brought her on to his team to take the lead on the behavioral component of the work. They have worked together ever since.

One of the best studied of all primate species, rhesus macaques can live for thirty years. Males tend to be a bit taller than females, and they tip the scales at about 8 kilograms while females weigh in around 5 kilograms. The macaques on Cayo Santiago can be a handful to work with, in part because of the large size of their social groups. The monkeys are often moving, and even when not, the tattoos and ear-notch marks they have been tagged with may be hidden by the bush or obscured by the angle between monkey and researcher. What that means is that Brent and others working with macaques not only have to maintain records on every monkey's fur color and pattern, eye features, and other traits that can help distinguish the island's residents from one another, but they must also memorize that information and be able to make split-second identifications among a group that can number from a few dozen to more than a hundred individuals. In time Brent mastered this skill, and when she did, she came to learn that the rhesus macaques' social networks impact so much about their life (and death) on Cayo.

One of Brent's early studies focused on females in two groups of macaques, one of which had fifty-eight females and the other about twenty. She'd select a macaque from her list, find it, follow it, record what it was doing, and then select another monkey and repeat the process. Macaques groom one another often, which serves not only to remove parasites from the recipient but to lower stress levels in all parties. Brent was especially interested in grooming networks in females, as well as networks based on the proximity of individuals when they were not grooming. She wanted to see what the networks looked like and whether they changed over time: in particular, whether network structure changed between the mating season in late March–late July, when females engage in sexual interactions with males, and the November–December birthing season, when females provide care for infants. To make her comparisons, she used network metrics like density—a measure of how tightly connected members of a group are—something that we will delve into in more detail in the following chapters.

Brent discovered that females were more tightly connected to one another, both in terms of grooming and proximity, during the mating season than the birthing season. Probing the network for information on key individuals, she found that during the mating season, but not in the birthing season, grooming and proximity networks revolved around a select few females. All of that made sense, because females are laser focused on their own infants during the birthing season.[4]

This was one of the earliest studies of how animal social networks change across seasons, and Brent saw it as just the first step. Now, with a sense of the basic network structure and dynamics, she and her colleagues could dig deeper. Assuming, of course, that the monkeys stopped stealing their iPads. Camille Testard, a graduate student working with Brent and Platt, was a victim of one such heist. She put her bag down for just a moment, but that was all it took. After the macaque thief ran up a tree, Testard caught up with him, and when she startled the bandit, he dropped the iPad into her waiting arms.

Monkey mischief aside, when Brent and her team looked at the data from 2010 to 2017, they found that females who had strong friendships (tight connections) with their favored partners in a network had higher survival probabilities than other females. Another way for a female to increase her chances of survival was through *weaker* connections but with *lots* of partners. It wasn't only the number of friends or the strength of specific favored friendships: friends of friends also matter to female macaques. The more friends of friends a female has, the more offspring she produces.[5]

By 2017 Brent and her colleagues—including research assistants like Daniel Phillips, who was on Cayo Santiago every day—had a deep understanding of the way that the social networks of macaques on Cayo worked and what they meant. There was, of course, always more to learn, but much of that would likely involve filling in the details. Then Hurricane Maria struck, devastating everything in its path, including the social networks the macaques were rooted in.

When Hurricane Maria hit Cayo full force, Brent's first thoughts were of the staff and research assistants, like Phillips, who live near Punta Santiago year-round. "[I] couldn't get in touch because all the cell towers were down. . . . No one heard from Danny for two, maybe three weeks after Hurricane Maria. . . . It was pretty horrible." Eventually though, she learned that Phillips and everyone on the project were safe. Then her thoughts turned to the macaques and the island itself. "When we saw the satellite tracker . . . I just thought, 'This field site is gone; these monkeys are all dead. . . . It's a Category 4 hurricane, and these animals were just sitting on the ground on this little chunk of rock in the middle of the ocean.'"

They weren't all dead, but Hurricane Maria turned the monkeys' world upside down and meant they needed to reestablish their society, with all its intricate and complex working parts. And quickly. For Brent and her team, Hurricane Maria meant many things, including figuring out new ways to think about the effects of large-scale natural disasters on social networks. For that they would, among other things, need to build on the work of some of the pioneers in the field of social network analysis in nonhumans. One of those pioneers was animal behaviorist David McDonald.

The tropical montane forests of Monteverde, Costa Rica, where David McDonald studied long-tailed manakins (*Chiroxiphia linearis*) are a sight to behold. Massive trees overgrown with strangler vines provide food for monkeys, while the understory is replete with shrubs of every kind along with trees from the coffee family. At any given moment, you might see flying overhead a beautiful resplendent quetzal (*Pharomachrus mocinno*), with its red underside and blue tail, or a three-wattled bellbird (*Procnias tricarunculatus*), half white and half chestnut brown with wattles that make it appear mustached.

For the better part of two decades, sitting in a birder's blind made of black plastic tarp thrown over a little wooden frame, McDonald was right in the middle of all that, year in and year out. He and his colleagues and assistants have logged more than 15,000 hours sitting in those blinds. McDonald and his team were there, in part, to watch a song-and-dance routine that would be the envy of any Broadway choreographer. Clusters of five to fifteen long-tailed manakin males—replete in their red, blue, and black plumage—spend their time in areas with perches from which they perform their routines to entice females into mating. Outside of the breeding season in April and May, any male in a cluster can practice singing and dancing, and they do just that. Even during the breeding season, every male is free to sing and dance in a perch zone—as long as there are no females around. When females are about, only the alpha and beta males—the two highest-ranking males in a group—can sing and dance at a perch. All other males are chased from perches by the performing pair. Once that is taken care of, the alpha and beta males—sitting on a well-defined perch and draped in their gorgeous plumage—put on their show for females.

McDonald's field site sits near the stunning Monteverde Cloud Forest. Much of the day, that forest is out of sight from inside the blind. "Observing the birds from a little cramped blind 10 meters away," McDonald says, "is a lot like . . . trench warfare in World War I. Long periods of utter boredom interspersed with short periods of sheer terror." That terror often begins when the alpha male emits a *teeamoo* call that attracts the beta male to the area. Then from up in the canopy, the pair begins a bout of well-synchronized, coordinated singing, producing a call that strikes the human ear—at least the ear of English speakers—as the word *toledo* (*toe-lee-doe*), with the first syllable in F-flat, the middle syllable rising to A-flat, and the final syllable again in F-flat. This is usually enough to lure in a female, but on occasion, an *owng* (*oh-wing*) call is tacked on.[6]

McDonald spent a lot of time counting the number of *toledo* calls. "They can do a million of those [*toledo* calls] in a season," he says. He used what he taped on handheld Sony cassette recorders—his PhD work on this system was done in the early 1980s when cassette recorders were state of the art—for subsequent analysis of the micro-details of songs. After the *toledo* calls attract a female, the males descend to a common perch, most often a branch or vine about 2 meters from the forest floor.

Once on the perch, the coordinated dancing begins, with an occasional call directed at females to make sure they are paying attention; though once the dancing gears up, it is hard to imagine a female paying attention to much else. While facing the female, males begin bouts of hops, leapfrogging over one another. The male closest to the female flies straight up about 60 centimeters, with its wings flapping at an eye-popping speed. The other male then slides toward the female as the bird in flight lands behind its leapfrogging partner. Males alternate roles, and a bout of hopping can involve up to 200 leapfrogging displays that are, on occasion, broken up by a "lightning drop" in which a male descends more precipitously than normal.

As the leapfrogging continues, the alpha male will sometimes emit *buzz-weent* calls toward the beta male. That causes the leapfrogging to speed up, and the dominant male may give off a *weent* call, lacking the *buzz* component, which leads to the beta male ending his dance. The dominant male then commences "butterflying," a type of solo flight where he circles over the perch, with the female watching intently. Mating between the dominant male and the female often takes place after that.[7]

Year after year, McDonald and his colleagues, including animal behaviorist Jill Trainer, went down to Costa Rica and banded long-tailed manakins with unique color-coded tags to identify the birds. The researchers spent endless hours in those blinds, taking notes on everything the manakins did, while their tape recorders gathered data on vocalizations. When not out in the forest, they lodged nearby in a Quaker community, renting whatever space was available that particular year.

It didn't take long for McDonald to learn that during the mating season, *nothing* gets in the way of males performing their songs and dances. One year there was an earthquake causing a forest fire that burned right up to his study site. "I fell to the ground, because of [the] earthquake," McDonald describes, "and I thought maybe I'd gotten smoke inhalation." Later, when he was talking to the farmer who owned the land where he was doing his research, McDonald couldn't help but smile when that farmer said the thing that impressed him most during the earthquake and fire incident was that the long-tailed manakins just kept on singing and dancing.

McDonald's study site contained many perch zones, and each had its own resident pair of alpha and beta males. But most males across the study site, even most older males, were not alphas or betas. McDonald soon became obsessed with understanding just who landed those two coveted roles. The first piece of that puzzle came when he learned that manakins are long-lived. Most species of small birds aren't, with many living just three or four years. Manakins are an exception (though we don't know why), weighing just half an ounce, "so you could almost mail two of them to grandma with one postage stamp," McDonald says, but "they can live twenty years."

The average age of a male manakin who mates is ten years old. It takes about eight years to reach the status of beta, and even longer to become an alpha male. An alpha-beta partnership can take years to form, but once it does, a pair often remains together for many more years. And that really matters, as McDonald and his colleagues discovered, because that synchronization improves as the pair stays together, and the more synchronized the *toledo* calls of a pair of alpha and beta males, the more females visit their perch; and the more females visiting a perch, the more matings at that perch. Almost all matings go to the alpha male, who can copulate with dozens of females in a breeding season. The beta male's apprentice role eventually reaps rewards after he inherits the perch zone, and all the benefits associated with it, when the older alpha dies or can no longer dominate him.

Long life and the deep social bonds that form along the way, thought McDonald, seemed to be at the heart of who ended up as alpha and beta males. But how did it work? He was certain that what *young* males did mattered when it came to achieving alpha or beta rank later in life, but even after studying the birds for more than fifteen years, he didn't have the theoretical tools to piece it all together. That changed in 2005, when McDonald was scanning articles in journals including *Nature* and *Science*. There he read about how sociologists were using social network analysis to study the dynamics of complex human societies, and McDonald thought that might just provide him with the tool he needed to understand life on the manakin perches. As he read more in depth, McDonald began to understand the way in which social network analysis works, mapping direct and indirect links between members of a group, following the flow of information between members, describing how well integrated a group of individuals is, and discerning who the central players driving a network are. Soon he was convinced this approach would give him what he'd need to answer the question about who ends up as alphas and betas.

McDonald knew that with the average age of a successfully breeding male being about ten years, a male would have had many other male associates during his formative years from ages one to six. Using 9,288 hours of data gathered from behind his blinds, McDonald looked at how many other males an individual was associating with during his formative years, as well as how many bonds those associates had formed with others around them. Then he constructed social networks to see if he could detect patterns in which males ended up as successful duet singers.

What McDonald discovered was that males most deeply embedded in their social network when young were the ones defending perches and performing duets later in life. To understand why, recall that although each perch zone has only two males performing duets for females, lots of other males are hanging out in a zone, practicing their performances when no females are nearby. McDonald found that networks that formed between ages one to six, when males were moving about between perch zones filled with other males, were the key predictor of which males would become alphas and betas about five years later. The greater a male's connectivity with other males at perch zones during those early years, the more likely he'd rise to the status of dueter, often with a friend (the male he was most connected to) or with a friend of a friend: every additional unit of connectivity in the network increased a male's chances of reaching alpha or beta status fivefold. During their formative years, males were building contacts, creating and solidifying their role in a social network in ways that paid off many years later.[8]

After his network research was published, McDonald was looking back at some files he had from his earliest days studying long-tailed manakins. He stumbled across an old diagram he had made but long forgotten: he had sketched a manakin network, with links between males, long before he had the tools to know what to do with it—he had been thinking about social networks all along, he just didn't know it.

When T. Brandt Ryder began his work studying the life and times of wire-tailed manakins (*Pipra filicauda*) in the Yasuní Biosphere Reserve in Ecuador, he knew all about social networks. Before he began that work, Ryder had read McDonald's social network studies on long-tailed manakins and had even recruited McDonald as a member of his PhD committee.

All fifty-three species of manakins, including wire-tailed and long-tailed manakins, engage in exaggerated courtship displays, but that fact masks important differences in how often those displays are coordinated. In some manakin species, displays are a one-bird show, with males singing and dancing in their territory: any male-male interactions are largely antagonistic. In other species, like McDonald's long-tailed manakins, displays almost always involve alpha and beta males performing coordinated and cooperative song-and-dance routines, to lure in females. And then there are species like the wire-tailed manakins, in which males sometimes show coordinated display behavior and sometimes go it alone: a system that may be an evolutionary step on the path from solo to obligate pairwise cooperative displays. For Ryder, this meant an opportunity to use social networks to dig into an understudied, but potentially important, stage in the evolution of cooperation.

Three hundred kilometers southeast of Quito, situated in a lowland Amazonian rainforest, the 744-hectare UNESCO Tiputini Biodiversity Station that Ryder called home while gathering his dissertation data was built as the result of an international collaboration between Universidad San Francisco de Quito and Boston University. The station sits in the 1.5 million-hectare Yasuní Biosphere Reserve, one of the most diverse ecosystems (per square meter) on the planet. Tiputini and the Yasuní Biosphere Reserve are home to 200 species of mammals, including giant anteaters (*Myrmecophaga tridactyla*), golden-mantled tamarins (*Saguinus tripartitus*), and dusky titi monkeys (*Callicebus discolor*); 150 species of amphibians, including map frogs (*Hypsiboas geographicus*) and reticulated poison frogs (*Ranitomeya ventrimaculata*); 250 fish species; and an astonishing 610 bird species, including Andean condors (*Vultur gryphus*), sparkling violetears (*Colibri coruscans*), giant hummingbirds (*Patagona gigas*), and wire-tailed manakins.[9]

Cooperative courtship displays in wire-tailed manakin males make up only 30% of all displays—the rest are solo affairs. When these displays are cooperative, they are as magnificent as the surroundings in which they take place. From a display perch, a pair of males that would barely register on a postage scale—each clad in gorgeous yellow, red, and black feathers like their long-tailed cousins—take turns in a synchronized eye-dazzling display for females that includes side-to-side jumps, back-and-forth jumps, sidling backward with their tail feathers raised, rapid hovering, as well as "swoop-in-flights" that entail flying up at full speed and then making an S-shaped turn over their partners.

Sitting for long stretches inside his birders' blind, binoculars and tape recorder in hand, Ryder took in all things wire-tailed manakin. Much of the time, manakins were not displaying, which gave him ample time to think about the function of those song-and-dance acts when they inevitably did come onstage. The problem was that it also gave him time to think about the endless attack of mosquitos, ants, and particularly sweat bees, which loved nothing more than a stationary, perspiring human target. Equally frustrating, but also part and parcel for fieldwork, were the times when an over-the-top coordinated display would unfold, and though he was able to discern the leg band, and so the identity, of one of the manakins, the leg band on the other was hidden by a leaf or some such thing. "Soul crushing" was how Ryder describes such turns of fate.[10]

Using information from 414 hours of observations, Ryder began by constructing social networks for males from three groups, largely based on coordinated, cooperative displays between males. At the most general level, in all three wire-tailed male networks, the average number of connections per bird was relatively low. But there was lots of variation between birds in their number of connections: most interactions in a network were driven by a small number of territory holders, acting as network hubs, and the males they interacted with.

A wire-tailed male's tenure on his territory was far and

away the best predictor of his reproductive success: a male's chances of fathering chicks increased fourfold for each additional year he held a territory. This creates intense pressure on males to eventually secure a territory, and the way they do that is by networking. To predict whether a male will rise from the status of non-territory holder to territory holder, what you need to know is the number of network connections he had: each additional partner that an individual engaged with in cooperative displays made it seven times more likely he would eventually land a territory.[11]

In a surprising twist, Ryder and animal behaviorist Roslyn Dakin found that the greater the average number of display partners that males in a network had, the less stable that network was over time. Most models suggest that the more deeply connected members of a network are, the more stable a network should be, which makes intuitive sense, because it should take more to disrupt sociality among friends. But the wire-tailed manakins were showing the opposite pattern. Why? At first, Ryder and Dakin were shaking their heads, searching for an answer. Then, they finally figured out what was going on: as individual birds interacted with more and more others, the potential pool of possible partners became much richer. All of a sudden, a male might discover a display partner he likes even better than the one he has. In time, all this shakes out, but as it does, the network goes through a period of instability along the way.[12]

The social networks in place in long-tailed and wire-tailed manakins demonstrate that networking is important not only in species where all displays are coordinated (long-tailed manakins), but in species where the majority of displays are one-male shows, and only about three in ten displays are coordinated affairs (as in the wire-tailed manakins): it may very well be that social networking is, in fact, a prerequisite for moving from lots of male-male aggression and occasional co-ordinated displays to forming bonds with more display partners and reaping the potentially greater payoffs for doing so.

Compared to the macaques on the island of Cayo Santiago, the long-tailed and wire-tailed manakins have gotten off easy, as their social networks have not been torn asunder by a Category 4 hurricane. So, what in fact happened to the networks of those highly social little macaque dudes and dudettes we learned of earlier?

More than 60% of the green vegetation on Cayo Santiago, along with a lot of the infrastructure that the Cayo team had put in place—including the feeders that supplied the animals with some of their food, which they now needed more than ever—was decimated by Hurricane Maria. But not a single macaque was killed during the hurricane, and only about 2% of the animals died shortly thereafter, probably due to starvation. "It's completely incredible," Lauren Brent says. "They're not that big, right? And all their trees were being blown over. It's not like you can hold on to something." At first Brent thought that maybe the macaques hid in a place known as Happy Valley, which is partially protected from wind, but normally Happy Valley holds fifty macaques, and there was no way it could have provided shelter for 1,700 monkeys. She and her team are working hard to piece together what happened, but still have not been able to figure out how every single macaque survived the full brunt of a Category 4 hurricane.

About three months after Maria, when the shock had partly worn off, Brent began thinking seriously again about the dynamics of macaque social networks, primarily because of what she was hearing from Danny Phillips and the other field assistants who were back on the island. They'd tell her "the monkeys are acting weird," and when Brent would ask how so, the on-the-ground team told her that they seemed to be especially friendly toward one another. In early 2018, about five months after the hurricane, she went down to check for herself. The island was still reeling from Maria (as was most of Puerto Rico). "But when I got to Cayo," Brent says, "I was like 'Yeah, I see it.' . . . They look more tolerant. . . . Monkeys I never expected to be cool just sitting next to each other.'"

At the time, very little was known about how animals adjusted their social dynamics after full-scale natural catastrophes, and as awful as the consequences of Hurricane Maria were for Puerto Rico, perhaps, Brent thought, one silver lining might be that the disaster shed light on those dynamics. When she and her team looked more deeply at her suddenly much more friendly macaques, what they found was that Maria had fundamentally altered the social networks of the monkeys. For one thing, when they compared the grooming and proximity networks in two groups of macaques during the three years prior to Maria versus the one year immediately after the hurricane, the data confirmed the anecdotal observations about the monkeys being nicer to each other: macaques were four times as likely to be found close to one another after the hurricane, and they were 50% more likely to groom one another. What's more, monkeys who groomed least and had spent the least time near others before Maria were the ones who showed the greatest increase in these behaviors after the hurricane.

Focusing their analysis on grooming behavior, Brent and her colleagues thought that the changes in network structure might be due to either an increase in the number of partners or an increase in time spent with specific partners. Or

Pre–Hurricane Maria

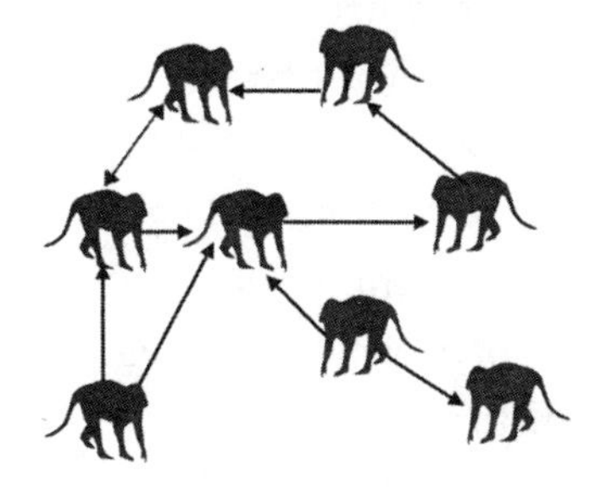

Post–Hurricane Maria

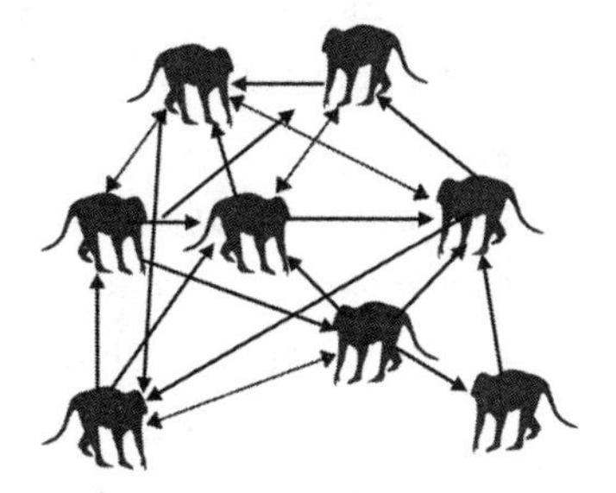

Grooming networks of rhesus macaque monkeys on the island of Cayo Santiago before and after Hurricane Maria. After Hurricane Maria, macaques had more social partners but had not formed stronger bonds with their previous partners. (C. Testard, M. Larson, M. M. Watowich, C. H. Kaplinsky, A. Bernau, M. Faulder, H. H. Marshall et al., "Rhesus Macaques Build New Social Connections after a Natural Disaster," *Current Biology* 31 (2021): 2299–309. Rhesus macaque image with permission from istock.com.

perhaps a bit of each. What they found was that post-Maria, macaques had more social partners, but the average strength of a grooming relationship with a partner had not changed. The monkeys had formed more friendships in their network, not strengthened already existing ones: Maria had brought the macaques in a group closer together, with additional grooming partners buffering them from the devastating effects that the hurricane left in its trail. And again, friends of friends mattered: monkeys took the path of least resistance in forming new relationships. If monkey 1 had been in a grooming relationship with monkey 2 before Maria, it was more likely to enter a grooming relationship with one of monkey 2's grooming partners after the storm.[13] Disaster, in the form of Maria, had brought the monkeys closer to one another, and social network analysis was the perfect means to show how.

With a sense of what animal social networks look like and what they mean to the animals who are part of them, we can now dig a bit deeper and look at the history of how the social network approach developed within the field of animal behavior, including how and why ethologists have imported many ideas from other disciplines and "evolutionized" them. Then, by examining social networks in everything from sulphur-crested cockatoos to manta rays and vervets, we will probe the theory behind social network analysis and get a finer sense of some of the theoretical nuts and bolts used by animal behaviorists studying networks.

2. The Ties That Bind

I am a firm believer that without speculation there is no good & original observation. — Charles Darwin to Alfred Russel Wallace, December 22, 1857

The seeds of social network thinking are sprinkled throughout the writings of Alfred Espinas. In June 1877, the thirty-three-year-old aspiring sociologist stood before the Faculty of Science at the Sorbonne to defend his doctoral dissertation, "Des sociétés animales." Espinas proposed that nonhuman societies could and should be studied by sociologists. Social behavior in animals, he argued, "constitutes the first chapter of sociology." It was a radical enough idea that some at the Sorbonne thought the topic was not science at all, but Espinas won that argument and was awarded his *doctorat ès lettres*. Animal societies, Espinas hypothesized, persisted over time and possessed group-level properties that arose from the complex web of social interactions between individuals. Other sociologists, like Raphaël Petrucci, were soon advocating similar positions.[1]

It took almost eight decades for these ideas to migrate to biology, in part because biologists interested in animals were unaware of work coming from sociology. But there was more to it than that. Early ethologists' focus was not on how group dynamics affected social behavior. Work in ethology in the early 1950s often cast social behaviors as innate and under the control of special "centers" within the brain. A stimulus acted as a key that unlocked a center in the brain that in turn caused the animal to act. Many researchers saw social organization as the sum of innate stimulus-response relationships. But some started to question these mechanistic explanations for behavior and initiated studies designed to help shed light on the role of social and ecological factors in structuring behavior within and between groups.[2]

By the late 1950s, animal behaviorists were building schematics of animal social groups that included not just what behaviors individuals were using, but the proximity of individuals to one another, group cohesiveness, group stability, whether a group was permeable (open to new members), and more. Soon ethologists were suggesting broadening the scope to include cliques—subgroups that act as a unit within which information flows from one member to another.[3]

Primatologists were the first on board, because many (mistakenly) assumed that the sort of complex dynamics inherent in this new approach required a primate-powered large brain and, more importantly, because some sociologists were studying (nonhuman) primates, so there was already some cross talk between sociology and primatology. Because of this, a few ethologists began using what was called the sociometric approach, developed to describe the structure of human groups and the positions of individuals within those groups. Sociometrics uses mathematical graph theory to describe the pattern of social relationships between people in a group. Individuals are represented as *nodes* and are connected by lines called *ties* (also called *edges*), that represent interactions such as aggression, trade, cooperation, among others. This sociometric approach was an important precursor to modern social network analysis.

In 1965, in a very early example of adapting the sociometric approach to nonhumans, primatologist Donald Sade constructed sociograms—what today we would call social networks—of grooming interactions among the ancestors of the rhesus macaques on Cayo Santiago that we learned of in the last chapter. Sade's sociograms captured not just grooming per se, but the strength of grooming relationships, by using *weighted ties*, where the thickness of a tie shows the relative frequency of grooming interactions: the thicker the line, the stronger the grooming relationship.[4]

In the mid-1970s, animal behaviorist Robert Hinde began to think more seriously about the notion that interactions between pairs of individuals could, in principle, influence the very dynamics of group life. Hinde realized that in sociometric models, nodes might just as easily represent individuals in *any* primate species, not just macaques: that, indeed, individuals in groups of birds, fish, insects, and more could be represented as nodes.[5]

But there were problems. For one thing, sociometric studies were hamstrung by a lack of computational power. Some of the measures used in sociometrics are simple to calculate by hand. It is easy enough to calculate a measure called *degree*, which simply counts up the number of individuals that an animal interacts with. *Strength*, measured by weighted ties, is just as easily calculated by using the relative frequency of interactions between pairs of animals in a network. But the use of many other measures, which we will delve into in a moment, had to await a greater availability of computing power.

When that computational power became readily available at the start of the twenty-first century, social network analysis

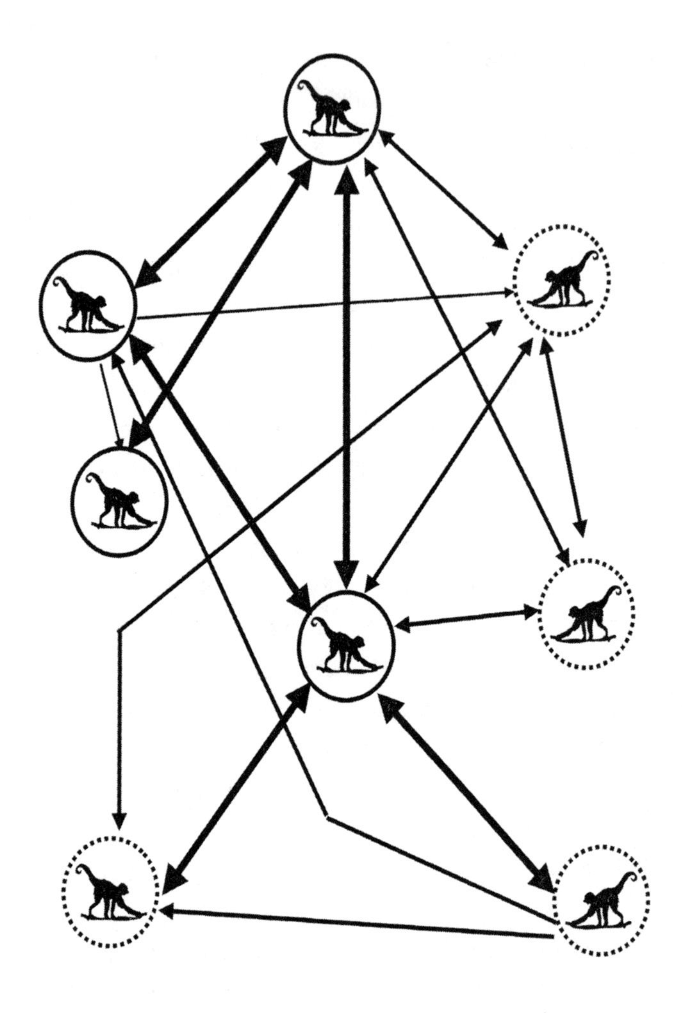

Grooming relationships in a group of rhesus macaques (solid circles = females; dashed circles = males). The *ties*—the lines connecting macaques—represent a grooming relationship. Ties here are *weighted*, and so the thickness of a line shows the relative strength of the grooming relationship. Ties here also show the directionality of grooming. Most of the grooming relationships run both ways, but some are unidirectional. Based on D. S. Sade, "Some Aspects of Parent-Offspring and Sibling Relations in a Group of Rhesus Monkeys, with a Discussion of Grooming," *American Journal of Physical Anthropology* 23 (1965): 1–17. Rhesus macaque image with permission from istock.com.

was born. With it came new measures that provided ethologists with the ability to dive much deeper into the complex underpinnings of networks in nonhumans. But social network analyses are more than simply high-powered versions of sociometrics. They seek to go beyond just describing network structure: these analyses are meant to elucidate how networks form, what properties they possess, and within the field of animal behavior, what their function is in both an ecological and evolutionary context. To see how they do this, and to dive into the social network tools that are now available to ethologists, we will explore manta ray networks in the breathtakingly beautiful waters of the Raja Ampat Regency of West Papua and social networks of vervets at the Mawana Game Reserve in South Africa. Before that, though, it's off to Sydney, Australia, to learn of a citizen science project on sulphur-crested cockatoo social networks.[6]

"When I was young," Lucy Aplin recalls, "hanging around the back corridors of the . . . [Western Australia] museum where Dad worked was basically my after-school day care. I have fantastic memories of being let loose amongst the exhibits." More than two decades later, Aplin was back at the museum, but this time as an up-and-coming ethologist on a research mission.

Aplin had a sulphur-crested cockatoo (*Cacatua galerita*) as a pet when she was young, and she had also seen them flying around her neighborhood, so she knew just how tame the birds were. She was reminded of that again when she was sitting with her brother on the balcony of his apartment near the Royal Botanic Garden Sydney. In 2015 Aplin was on holiday from a Junior Fellowship at Oxford University, where she was studying social networks, when she found herself hand-feeding sulphur-crested cockatoos that had stopped by her brother's flat, as she learned they did quite often. She noticed that some of these birds had a numbered tag on their wings, and her brother was using his smartphone to report which birds were there with an app called Wingtags. "I thought this was incredible," Aplin says. "You could do so much with social networks."

The brainchild of John Martin and Richard Major at the Australian Museum, Wingtags is a citizen science project designed to pair Sydney residents' love for cockatoos with a means for gathering behavioral and ecological data on the birds. The cockatoos like to hang around the Royal Botanic Garden, with its thousands of exotic and native plants, including native trees with hollows that cockatoos use for nesting and roosting, and so Martin, Major, and their team captured 136 sulphur-crested cockatoos there and placed a numbered tag on each bird's wing. The vast majority of tagged birds stay within a 10-kilometer radius of the gardens.[7]

The Wingtags app went live in 2012, and since then it and its offshoot, an app called Big City Birds, have received more than 27,000 reports from nearly a thousand residents of Sydney. When a bird is sighted, a citizen scientist snaps a picture that includes the cockatoo's wing tag band and uploads the image, and the app then notes the GPS location of the bird. After more than 8,000 images were examined by volunteers, it became clear that most birds were feeding when observed by Wingtag users.

Aplin went to the museum to talk with Major about returning for a six-month visit one day to use Wingtags data to construct and analyze cockatoo foraging networks. When she arrived in 2016, which by chance was also the 200th anniversary of the Royal Botanic Garden, Aplin spent virtually all of her time on two tasks. For one, she sifted through the reams of digital data that Wingtags had amassed since it launched, trying to figure out how to translate that data into a form that could be used to study the birds' social networks. From discussions with Major and others, and from her own knowledge of cockatoo behavior, Aplin decided that if the GPS coordinates showed cockatoos within 100 meters of one another during a thirty-minute period, they would be considered part of the same group for that time frame.

The second task was trickier. If she was going to build a social network, Aplin wanted it based on a representative sampling of the tagged cockatoos. "Maybe . . . people are just reporting particular [cockatoos] they like," she thought, "and not other individuals," which is not as far-fetched as it seems, as Aplin knew that people form strong bonds with these birds. "It's completely unlike [what] you're familiar with in Europe and North America," she says. "People really individually recognize these birds . . . and they're convinced that . . . the birds recognize them." Certainly, one citizen scientist and cockatoo number 125 had formed a special friendship, as Aplin learned one day when she got an email from this fellow who had been feeding the cockatoo for years. "Every day on the dot, it turns up on my window at my kitchen," he wrote Aplin. "I leave the window open, it walks in, and we share breakfast on the table. . . . This bird and me have such a close relationship that my wife has been calling it [my] girlfriend." Most of the years that cockatoo 125 was visiting were before the Wingtags group caught and tagged it. "Imagine my surprise," the emailer wrote, "when this morning it turned up with this big plastic tag [on its wing]." What he wrote next surprised and amused Aplin: "You have my permission to continue [with this work], but can I name the bird? . . . Can you please call it 'Girlfriend'?"

To make sure that the Wingtags' data was indeed unbiased, Aplin captured nine of the wing-tagged cockatoos and fitted them with solar-powered 10-gram GPS tags that, under ideal conditions, could gather data on the birds' locations once every second. Then, carrying a transmitter that could download the data from the GPS tags, she would head to the Royal Botanic Garden and a few other locales that the birds frequented. Once she had downloaded data from the GPS tags on those nine birds, Aplin compared it with the GPS data collected when citizen scientists uploaded their photos of the same birds for the same period, and she found that the citizen science data was indeed similar to her own, and so likely unbiased.

Sifting through the data on nearly all wing-tagged cockatoos—which had been seen, on average, 147 times each—Aplin and her colleagues used group membership to construct foraging networks each year for six years. They then looked at numerous network metrics to better understand what drove foraging network structure in the cockatoos, including the number of individuals a bird associated with and how frequently a given pair of cockatoos interacted.

Aplin and her team found that, on average, younger birds formed stronger bonds with more partners than did older birds. As individuals grew older, the strength of the bonds they formed became more and more consistent over time. Season also affected foraging networks. Though cockatoos didn't associate with more of the members of their network in the winter months (March–August) when food is harder to come by, they strengthened the bonds with the network associates they already had.

Colder, harsher times also forced cockatoos to change their networking habits in a more dramatic way. To see how, we need to introduce a social network metric called the *clustering coefficient*, which measures the extent to which a cockatoo's neighbors are connected to one another. When the

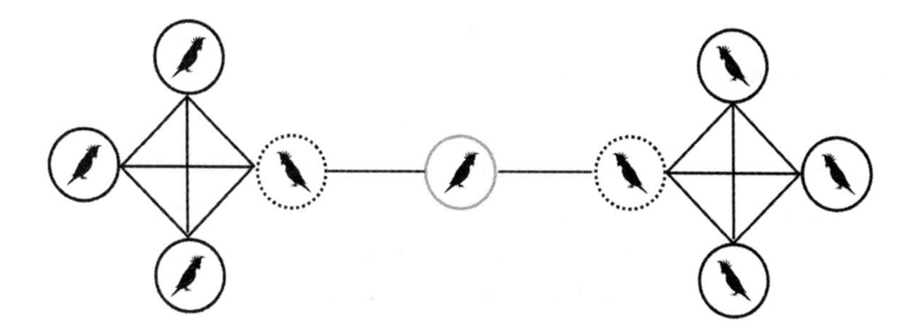

A schematic of clustering in a social network. In this simple, hypothetical network, clustering measures the extent to which a sulphur-crested cockatoo's network neighbors are well connected to one another. Birds in solid black circles have high clustering values, as each of their three neighbors are connected to one another. Birds in dashed black circles have moderate clustering values as some of their neighbors are connected to one another. The bird in the gray circle in the middle has a clustering value of zero, as its neighbors (in dashed circles) are not connected to one another at all, except through it. At the level of the network, two cliques are obvious. Technically, this measure of clustering is called the clustering coefficient. Cockatoo image with permission from istock.com.

clustering coefficient is averaged over all the birds in a social network, it provides a measure of how cliquish a network is. In Aplin's cockatoos, the clustering coefficient at the network level increased in the winter, suggesting the birds deal with hard times by moving about in search of food in tightly knit cliques.[8]

Big City Birds, which followed Wingtags, allows users not only to upload an image, but to provide a detailed report on what a bird, or group of birds, is doing, above and beyond foraging. Citizen scientists can now append notes about aggressive behavior, courtship, and other behaviors. In time, this will allow Aplin and her team to use strength and clustering to compare and contrast cockatoo networks across a wide range of social behaviors and, in so doing, get an even more nuanced understanding of the complex web of interactions in their networked society.

Strength and clustering are just two ways to measure what's happening in an animal social network. There are others, but we before delve into them, a bit more on the importance of strength scores, because for the rock hyraxes (*Procavia capensis*) who roam the Ein Gedi Nature Reserve, it's not so much about an individual's strength score as it is the *spread* of strength scores across hyraxes in a network. And that spread has implications for life and death.

Amiyaal Ilany and his colleagues have been studying rock hyraxes—small, furry animals weighing in at 3 kilograms and standing 25 centimeters tall, but whose closest living relatives are elephants and manatees—in a quintessential desert oasis in the Ein Gedi Nature Reserve. Stunning waterfalls at Ein Gedi cascade down cliffsides into springs that provide the hyraxes, Nubian ibexes (*Capra nubiana*), fan-tailed ravens (*Corvus rhipidurus*), sand partridges (*Ammoperdix heyi*), and other animals with the lifeblood to survive in an environment that often tops 35°C at midday.

The rock hyraxes at Ein Gedi live in colonies on rock outcroppings that pepper the cliffs in the reserve. They are a noisy lot. Males sing in bouts that can last many minutes and are chock-full of wails, chucks, snorts, squeaks, and tweets. Ilany's first dive into the social lives of the hyraxes was his PhD research on those songs. But the more time he spent at Ein Gedi with the hyraxes, the more fascinated he became with their other social behaviors. And Ein Gedi was also a special place to Ilany for a more personal reason. His father had been one of the first zoologists in Israel and decades earlier had done long-term work on leopards (*Panthera pardus*), long since gone, in the area around Ein Gedi.[9]

The challenge for Ilany was how to map out all the hyrax social interactions he was observing. Then one day he went to theoretical ecologist Yael Artzy-Randrup's PhD defense, where she talked about social network models. "It hit me," Ilany says, "a real Eureka moment." He read everything he could get his hands on about social network modeling, and in time he did two postdoctoral fellowships, where he dove even deeper into the topic. Then he went to the field to apply network theory to rock hyrax sociality.

Ilany's field season at Ein Gedi starts in March and lasts about five months a year. It's physically taxing work. He gets up at 3 a.m. to climb out of either the David or Arugot Gorge to reach the hyraxes before they become active for a few hours until the midday sun takes them to their burrows, where they stay until late afternoon, when they reemerge for a few more hours. Ilany finds a spot no more than 50 meters away from the hyraxes and sits with binoculars and a spotting scope, gathering data on social interactions. Each adult hyrax has a collar with a unique identification: juveniles can't be collared as they might choke as they grow, and so Ilany uses color-coded earrings for younger animals.

To look at social networks, Ilany used 1,500 hours of observations that he and his colleagues had made of two hyrax populations, focusing on which hyraxes travel together in tight unison around the cliffs as well as hyrax huddling behavior, both during the day and when the animals sleep in their burrows. Networking mattered. A lot. Indeed, network dynamics predicted how long a hyrax lived.

Ilany and his colleagues calculated the strength score for each hyrax in their networks. But it wasn't strength at the individual level—measured by how many connections a hyrax had and how strong the connections were—that mattered when it came to longevity. Instead, it was the spread around the average strength scores in a network that was key. In egalitarian hyrax networks, where strength scores were similar between individuals in a network, longevity was greater than in networks where there were a few central individuals with especially high strength scores. When it comes to cooperative behaviors like huddling, sleeping together in a burrow, and more, life in egalitarian societies trumps life in networks with the privileged few.[10]

On the northwest tip of West Papua, far from hyraxes and sulphur-crested cockatoos, Rob Perryman has been using a social network measure called betweenness to help him understand social behavior in the reef manta rays (*Mobula alfredi*) that were front and center in his PhD project.

Perryman earned his PhD in 2020, but his fascination with the reef manta rays began about a decade earlier. After watching documentaries on marine life when he was a child, he developed a passion for these majestic animals. Soon Perryman was taking a diving course in Mozambique, because he knew the waters there were full of rays. After that dive course, where he met quite a few manta rays up close and personal, he wanted to know more. "The thing that struck me most when I was in the water with [the rays] is how curious they are. . . . They also seemed to be social with each other as well." To learn more about just how social, he enrolled in the European Union's Erasmus Programme, where students can do a master's degree research project anywhere on the planet on whatever topic they are passionate about.

This time, rather than Mozambique, Perryman decided he would study manta rays in the Dampier Strait in the Raja Ampat area of West Papua, where they tend to be found in shallow waters, which would make diving all the easier (and safer). What's more, the area has a thriving tourist trade where people come to watch the rays in some of the most beautiful waters on the planet. This meant that Perryman might be able to hitch a ride on a tourist boat, put on his scuba suit, roll off the deck with an underwater camera or video recorder in hand, and watch his subjects.

The rays Perryman studied frequent "cleaning stations," usually found on a boulder covered in coral, where small reef fish like Klein's butterflyfish (*Chaetodon kleinii*) will pick parasites off a ray's body and even swim into a ray's mouth to remove bits of plankton that may be stuck there. During the day, the rays also visit naturally occurring "feeding stations," often in the shallow waters of the reef, feasting on various species of plankton, worms, shrimp, and the like.

Perryman's work in the Dampier Strait began on a speck of an island called Arborek: the population of rays that swim in the waters around Arborek far outnumbered the 130 or so residents of the island. It didn't take long before Perryman discovered that if he wanted to observe the rays interacting in groups, the two best places to do that were at feeding and cleaning stations, and so he remained on board until the boat he hitched a ride on that day approached one of those spots.

Each day Perryman would do a pair of hour-long dives. He'd usually just head to the sandy bottom and sit himself down, as the rays were less spooked by that than by a human swimming around. There he would take still shots or record the rays on an underwater video camera, trying to get as much information as he could about their behavior. He could recognize rays based on the photographs and videos because each ray has a unique spot pattern on its upper surface. In between dives, he took advantage of the fact that the rays spent about a third of their day swimming seemingly effortlessly at the water's surface, and he flew a drone he had brought with him to get some footage from above.

Perryman yearned to be testing hypotheses about manta ray social behavior. Over the next two years, he went back to Arborek a few more times, using his own funds, so he could be better prepared to enter a PhD program with some well-formulated hypotheses. During that time, following up on a suggestion by a fellow student in the Erasmus Programme, he also began a deep dive of another sort, this one into the literature on social networks in nonhumans. When he came up for air from that mathematical plunge, he had the conceptual framework he needed.

For his PhD, Perryman worked with Culum Brown, who had studied social networks in both sharks and dolphins and was happy to have a new student look at social networks in another exciting system. Now, rather than using Arborek as his home base, Perryman stayed on the nearby larger island of Gam, one of the world's premier dive destinations, where wealthy tourists shell out good money to visit, dive, and stay at high-end resorts. Perryman and others worked out a deal, where he gave an occasional lecture on his research and marine life in general, and in return he got his room and board for free.[11]

Like on Arborek, he dove and used his drone to gather data in the waters near Gam, but now Perryman had another tool at his disposal, as he had placed acoustic tags on twenty-nine rays. The tags fall off on their own after a year or so, and they don't do any lasting harm to the rays. Once the tags were in place, Perryman put out acoustic receivers at key areas like feeding and cleaning stations, so he could monitor the long-term movement behavior of the tagged rays.

To construct a manta ray social network, Perryman used more than 3,000 observations of the almost 600 rays he had come to know at three feeding stations and two cleaning stations in Dampier Strait. One of the more interesting social

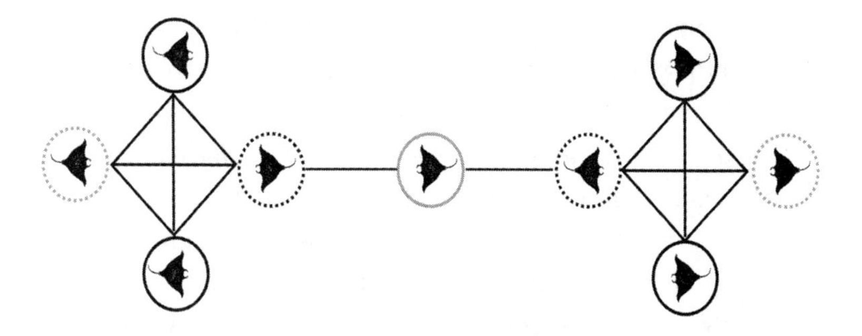

"Betweenness" in social networks. Individuals that have high betweenness link together many others in a network and have important effects on the flow of information, disease, or resources through a population. To calculate betweenness for a manta ray, select an individual and ask whether the shortest path between a pair of other manta rays in the network passes through that individual. If it does, add one to the tally, and then repeat the procedure with that same manta ray and every other pair of manta rays in the network. For example, to calculate the betweenness for the manta ray in the solid gray circle, begin by asking whether the shortest path between the two manta rays in dashed gray circles (far left and far right) passes through the manta ray in the solid gray circle. It does, so the betweenness tally is set at one. Then repeat this process for every pair of manta rays in the network, supplementing the tally by one whenever the manta ray in the solid gray circle lies on the shortest path between a pair. The manta ray in the solid gray circle has the highest betweenness in this network, the manta rays in dashed black circles have a slightly lower betweenness, and the manta rays in solid black circles and dashed gray circles have the lowest betweenness scores. Manta ray image with permission from istock.com.

network metrics that Perryman looked at was "betweenness," which measures how many of the shortest paths linking pairs of rays in a network run through a given individual. Individuals with high betweenness scores are key hubs in the flow of information and other resources through a network.

In the manta rays of the Dampier Strait, juveniles were the hubs, with the highest betweenness scores in a network.

Perryman is not certain why that is, but he thinks it may have something to do with juveniles being a bit less mobile than adults, and so once at a feeding or cleaning station, they are more likely to act as hubs. Conversely, females who had been pregnant at least once during the study had the lowest betweenness scores. "My thinking is that when they're pregnant, they prefer to socialize with a smaller number of . . . other females and try to avoid males," says Perryman. "We know males harass heavily pregnant females, following them and waiting for the baby to drop and for them to become sexually receptive again."

The social network analysis of rays also found two communities nested within the networks: one of these was heavily female biased, and other had a more even sex ratio. To see whether there might be differences in social dynamics between these two communities, Perryman used a metric called social differentiation, which looks at how much variation exists in the strength of bonds linking pairs of rays. When the sex ratio was female biased, the bonds linking members of the communities were strong and similar across all pairs and were stable over time: a tight, well-integrated community with members who were in it for the long haul. But in the community that had close to an even sex ratio, a more differentiated social structure was in place. Females in that community tended to be tightly connected, and though males were linked to others in the sub-network, the connections were weaker. Female or male, old or young, as rays gracefully glide through the stunning aquatic landscape of the Dampier Strait, they do so linked to one another in social networks.[12]

Vervet monkeys (*Chlorocebus pygerythrus*) in KwaZulu-Natal, South Africa, form social networks for nearly everything they do: play, groom, struggle for power, among other things.

Ethologist Charlotte Canteloup intended to leverage the vervets' predilection for network formation to understand the dynamics and complexity of the social life of monkeys living in six troops—the Noha, Baie Dankie, Ankhase, Kubu, Lemon Tree, and Crossing troops—at the 10,000-hectare Mawana Game Reserve in KwaZulu-Natal, South Africa.

Like every researcher who joins the long-standing Inkawu Vervet Project (IVP: *inkawu* is Zulu for monkey), when Canteloup first arrived at the Mawana Game Reserve in 2017, she needed to acquaint herself with the monkeys and with their surroundings, before she could start gathering data on social networks. For the first two months, she'd go out on her own each day and learn to recognize individual vervets and get a feel for their behavioral repertoire. Canteloup had photographs of each vervet that she could refer to, as well as descriptions of scars, fur pattern, and the like, but learning to associate all that with specific monkeys, so that she could make instant identifications, took practice. Spending eight hours a day, six days a week with the animals not only allowed Canteloup to get to know specific vervets, but to learn a bit about the idiosyncrasies of some of them. "Every individual in each group [is] different," Canteloup says. "At some point you know you should pay attention [to] this one [who] is a bit scary or aggressive or sneaky, and that one you can trust, no problem, because he's super nice." Those two months were also a lesson on the lay of the land. With venomous cobras and boomslange and the occasional honey badger lurking about, "you'd better know where you are and [how] to find your way," Canteloup says, "especially when you go alone in the bush."

The Noha troop provided Canteloup with a chance to study social networks to address some important but understudied questions about the dynamics of group life: How consistent is an animal's position across different social networks? If, for example, a vervet holds an especially important role in its grooming network, does it have a similar role within its play and power networks? What's more, Canteloup knew that juveniles were usually ignored in network analyses, and she thought that was a mistake. The Noha troop was home to between thirteen and nineteen juveniles, which meant it was perfect for adding juveniles to the social network mix.

Vervets in the Noha troop are a busy bunch, and so Canteloup and her team were also busy gathering data on almost 800 aggressive interactions, such as "chase," "hit," "bite," "grab," "lunge," and "steal food," to construct a power network, 7,000-plus bouts of grooming to build a grooming network, and 864 rounds of play between vervets to assemble a play network. To look at both consistency across networks and the role of juveniles in those networks, Canteloup and her colleagues used a measure called *social capital* (also referred to as *weighted centrality*). Social capital incorporates not just direct connections and indirect connections between network members, but also how strong those connections are (by taking into account the frequency of interactions).[13]

Canteloup and her colleagues used social capital to look at consistency across the two years of their study as well as across different types of social networks. For each of the networks—power, grooming, and play—social capital scores for vervets were similar across years. This was true even though troop membership in Noha changed across the years: despite having a slightly different pool of members to interact with, vervets that were part of the Noha troop both years had similar social capital scores in power, grooming, and play networks across years one and two.

Consistency in social capital *across* networks was more complicated. Vervets tended to have similar social capital scores in their grooming and power networks: monkeys who

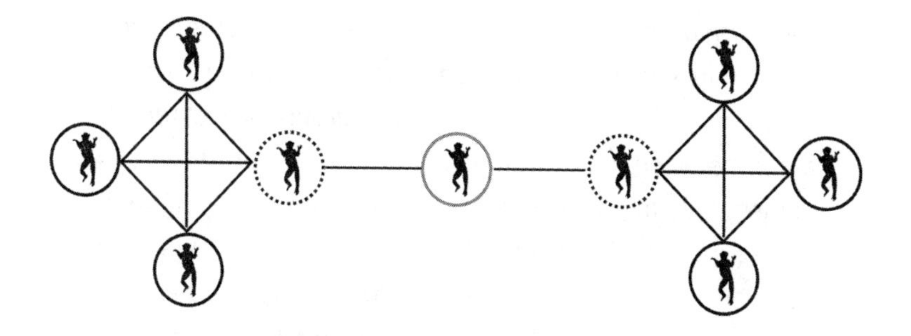

Centrality is a measure of how well connected an individual in a network is, looking at both direct connections with others in the network, and indirect connections, by taking into account how well connected that individual's neighbors are. In essence, it adds direct interactions onto the model of the clustering we looked at earlier. The vervet monkeys in dashed black circles have the highest centrality in this network, as they each have four neighbors, three of whom are connected to three other network members, and one who is not as well connected. Vervets in solid black circles also have high centrality, as they each have three neighbors, all of whom are connected to three or four other network members. The vervet in the gray circle has the lowest connectivity. Technically, there are a number of different measures of centrality: this one is called eigenvector centrality. Vervet image with permission from istock.com.

had high social capital in one had high social capital in the other. But social capital in the play network was not significantly correlated with social capital in either the grooming or power networks. And that's where incorporating juveniles into the social networks became very useful.

Juveniles played a role in grooming and power networks, but it was a relatively minor one, and they were on the wrong end of the hierarchy when it came to power. Play was another matter: while vervets of all ages play, juvenile males were the central actors in the play network. One reason play networks may be especially important for juvenile males is that when males reach sexual maturity, they disperse from their troop, often together, to find mates in another troop. Canteloup and her colleagues hypothesize that social capital invested in play networks in the Noha group may foster alliances between males who disperse as a unit when the time comes. Play in juvenile vervets also includes play fighting, and so juveniles may not only be forming nascent alliances but sharpening their fighting skills for the future when they encounter aggressive, and quite unwelcoming, males in the troop they attempt to join.[14]

Twenty years ago, very few animal behaviorists had ever heard of degree, social capital, centrality, clustering coefficients, and betweenness. Today, thanks to social network analyses, they know such metrics help shed light, at a whole new level, on the complex dynamics, the intricacies, and the wonders of animal societies.

3. The Food Network

Tell me what you eat, and I shall tell you who you are.
—Jean Anthelme Brillat-Savarin, *The Physiology of Taste* (1825)

David Lusseau always wanted to be a biologist. "Well, either biologist or clown," he adds, "but I realized there was not much money in clowning." When Marie the dolphin entered Lusseau's life, she sealed the deal for him becoming a biologist. A bottlenose dolphin (*Tursiops truncatus*) who swam in the waters near the village of Cerbère on the border between France and Spain in the late 1980s, Marie set seventeen-year-old Lusseau on a path that would one day lead him to study social networks in her species. "When you look in the eyes of a dolphin, you realize there is a lot going on," Lusseau says, reminiscing on his time with his cetacean friend. "It is something that is very hard to express or grasp or explain in a factual matter, but spending time with [Marie] got me interested in . . . trying to understand how dolphins work, [in what] I perceived as another intelligent species on the planet."

As an undergraduate, Lusseau spent time as a research assistant working with a group studying bottlenose dolphins in Florida. When out in the water, he encountered dolphins swimming on their own or in pairs. On occasion he bumped into a trio, but dolphins always seemed be doing their own thing, just in the company of one or two others. That view of dolphin sociality, or the lack of it, changed dramatically when Lusseau began his PhD research in the late 1990s at the University of Otago in New Zealand. His dissertation focused on conservation biology in bottlenose dolphins in a fjord called Doubtful Sound, but the social behavior of the dolphins there hit him like a ton of bricks. As soon as he got there, he encountered not lone dolphins, duos, or trios, but groups of thirty or more dolphins schooling and moving about in a coordinated manner. These were very different animals from the solitary dolphins and very small dolphin groups he had studied in Florida.

Each day Lusseau rose at 4 a.m., grabbed some breakfast, swatted away an endless barrage of midges, and arrived at Doubtful Sound before the sun rose. He'd board a 14-foot boat, locate a group of dolphins, and do focal animal sampling, cycling through dolphins, each recognizable by natural markings, often from shark attacks, on their dorsal fins. Doubtful Sound can be stunningly beautiful, but it is at a latitude called the "roaring forties" because of the strong winds from the west and six- to eight-foot waves at times, which make for rough going when watching dolphins from a boat.

As he spent time with the dolphins, Lusseau began thinking about how to understand their complex social dynamics, but he couldn't quite figure out the best way to proceed. On one of his stints back to the University of Otago, Lusseau recalls reading a *Proceedings of the National Academy of Sciences* paper on social networks written by physicist Mark Newman and others. Soon after that, he emailed Newman, telling him, "I think you are doing really cool stuff and I can understand it, because you write so well. Would you like to have a look at what we're doing?" Newman was interested. It wasn't long before he and Lusseau were coauthoring papers on dolphin social networks. But before they penned any coauthored papers, Lusseau published a 2003 paper in the *Proceedings of the Royal Society of London* that is widely regarded as the first study explicitly on social networks in nonhumans.[1]

Unlike animal social network papers in today's journals, where readers are acquainted with how networks operate, to put readers in the right frame of mind in 2003, Lusseau opened his *Royal Society* paper using a strategy that Darwin had employed in *On the Origin of Species*. The idea was to introduce a phenomenon that readers already knew about (in Darwin's case artificial selection, as in selection of different breeds of pigeons) and then make the case that what followed (natural selection), though it appeared radical, was really just another variety of what he had just discussed. In Lusseau's paper, the opening sentences read: "Complex networks that contain many members such as human societies . . . the World Wide Web (WWW) . . . or electric power grids . . . permit all components (or vertices) in the network to be linked by a short chain of intermediate vertices." And before readers knew it, they were learning about such social networks in dolphins.[2]

Lusseau constructed dolphin networks based on thousands of observations, and one metric he looked at was network diameter, which measures the average shortest path between nodes. To introduce network diameter to readers, Lusseau first discussed psychologist Stanley Milgram's "small world" research from the late 1960s. "The global human population seems to have a diameter of six," wrote Milgram, "meaning that any two humans can be linked using five intermediate acquaintances."[3] The party version of Milgram's small world is the parlor game "six degrees of Kevin Bacon." The rules are simple: players choose a movie actor and then connect that actor to another that they played alongside in a film, repeating the process over and over, trying to link their original actor to movie star Kevin Bacon—who once quipped "he had worked with everybody in Hollywood or someone who's worked with them"—in no more than six connections. It turns out the dolphin small world in Doubtful Sound is smaller than the human one (including Kevin Bacon's), both in the size of the network and network diameter, the latter of which is approximately three, meaning any two dolphins in Doubtful Sound can be linked using two intermediate acquaintances.

Lusseau wondered what would happen if the dolphin network was culled by, for example, shark predation. To do this, using the network data he had collected, he built a computer algorithm that simulated predation, reducing the network size 20% by randomly removing 20% of the dolphins. The small world of the dolphins, it turned out, was unaffected by such a reduction. But if instead of randomly selecting individuals to remove from the network, Lusseau simulated removal of the 20% of dolphins who had the greatest number of ties to others, network diameter increased, which had the effect of slowing information transfer within the network.[4]

As he came to know his dolphins better, Lusseau discovered that some individuals in Doubtful Sound give signals that affect group movement associated with finding new resources, including food. Side flopping, in which a dolphin leaps from the water and lands on its side, is seen only in males when they initiate a move to a new location. Upside-downing, in which an individual rolls onto its ventral side and slaps the water to signal an end to a group move, is seen almost exclusively in females. But only a few males do all the side flopping, and only a few females do all the upside-downing. Lusseau wanted to know if a network analysis would shed light on exactly which males and which females. It did. Males initiating and females terminating travel had higher betweenness—they were key hubs in this traveling/foraging network—than their nonsignaling counterparts.[5]

In a few populations of bottleneck dolphins on the other side of the planet, in Brazil, signaling and networking is not *sometimes* about feeding opportunities—they are *always* about that. And the dolphins there have, rather remarkably, added humans to their feeding networks.

Given how important foraging is in the day-to-day life of animals, it isn't surprising that natural selection has favored networked foraging in a myriad of species. Vampire bats regurgitate food to the hungry in their network, crop-raiding elephants network as they wreak havoc on farms, and black-capped chickadees are embedded in foraging networks, in both rural and urban settings. So, before we take a deeper dive into (and alongside) those dolphin/human networks in Brazil, let's sneak a look into some rotted-out trees and explore the vampire networks within.

A bad reputation is hard to shake, particularly if you are portrayed as the manifestation of evil. But that's the fate that Bram Stoker's *Dracula*—Romanian for devil—has bestowed upon the poor vampire bat (*Desmodus rotundus*). Fortunately, animal behaviorists are using the bats' own social behavior to show just how misguided the pop culture view is. That rehabilitation began in the early 1980s, when behavioral ecologist Gerald Wilkinson found that vampire bats cooperated with one another in potentially lifesaving ways.

The vampire bat roosts that Wilkinson studied in northwest Costa Rica were home to about a dozen females and their pups. Weighing in at between 25 and 40 grams, vampire bats do indeed rely on blood meals, largely from cattle and other domesticated species. The problem for vampire bats is that, compared to moving about on land, flying requires a lot of energy: so much so, that if a bat doesn't get a blood meal every second or third day, it can starve to death. There are two ways to get a blood meal: either fly out and clamp onto an unsuspecting cow or beg another vampire to regurgitate some of the blood meal in their gut directly into your mouth and hope they oblige, which on occasion they do. The begging route to a good blood meal is where a form of cooperation, called *reciprocal altruism*, comes into play.

In 1971 evolutionary biologist Robert Trivers hypothesized that under certain conditions natural selection will favor a strategy of reciprocal altruism, whereby two individuals exchange acts of cooperation and altruism. Trivers proposed that this sort of reciprocity would do especially well when animals live in groups and interact with the same partners, and when individuals can remember those who have helped them in the past. Vampire bats fit the bill perfectly. And when Wilkinson looked to see if bats exchanged acts of food sharing, he found some evidence that they do: bats were more likely to give blood to those who had donated blood meals to them in the past.[6]

Vampire bats are clearly capable of both cooperating and some rather sophisticated scorekeeping. Do these skills, among others, create a food-sharing network that is broader than pairwise interactions? Simon Ripperger, an animal behaviorist at the Museum für Naturkunde in Berlin, had cause to wonder. Ripperger was part of a bat research team that was funded to build an automated tracking system for studying social behavior. What they came up with was a next-generation hardware and software package with sensors that not only tracked a bat in real time, but that also "talked" to each other. These proximity sensors, weighing a shade under 2 grams, sent out signals every two seconds and were enclosed in a 3D printed casing with a transceiver that allowed the sensors to communicate with each other and with various base stations placed in a bat's environment. The default setting on a proximity sensor was low-power mode, but when another bat with a sensor came within 5 meters or so, the sensors on both individuals "woke up" and a "meeting" began. That meeting was over when no signal was picked up for ten seconds. At that point, information on the meeting participants, the time of the meeting and its duration, and the maximum strength of the signal, which was used as an estimate of just how close the bats were to each other, were stored in the sensor's memory, to be downloaded later.

Ripperger and his colleagues first tested their system on fringe-lipped bats (*Trachops cirrhosus*) near the Smithsonian Tropical Research Institute (STRI) in Gamboa, Panama, not far from the Panama Canal. It didn't go well. The hardware and software obliged, but the fringe-lipped bats did not. "The bats would just take off with the sensors," Ripperger says, "and then they'd never show up again." Fortunately, the STRI is a hub for bat ecologists and animal behaviorists, and on one visit Ripperger met Gerald Carter, a former PhD student of Wilkinson's, who was doing work on reciprocity and food sharing.

Carter had a long-term research project studying food sharing and grooming in vampire bats, but up to that point, his studies had been done in controlled laboratory settings. "[Carter] was always wondering whether what he's studying in captivity was real. . . . How would that happen in nature?" Ripperger recalls. It became apparent, very quickly, to both of them that the technology Ripperger had was perfect to study vampire bats in the wild. The talking sensors would provide the sort of fine-scale monitoring to study social dynamics in a roost, and therefore potentially shed light on the sharing of blood meals and on other behaviors. Those sensors would also allow a social network analysis of food-sharing and grooming behavior in a way that was simply not possible otherwise.

Ripperger had taken a course on social network modeling that computer scientist–turned–animal behaviorist Damien Farine offered at the University of Konstanz, and Carter had been a postdoctoral fellow of Farine's; both were itching to apply the social network skills they had learned.

When Ripperger joined the vampire bat project, Carter already had a colony of bats he was studying in the laboratories at the STRI. A year earlier, in late 2015, Carter traveled about 700 kilometers northeast to Tolé, an area full of cattle pastures that provided vampire bats with many a target for meals. There, he found a large hollow tree that was home to about 200 vampire bats. Carter captured 41 bats, and from those he established an experimental colony of 17 females and 6 of their female offspring in a large outdoor enclosure at the STRI, where they were fed on blood that a local slaughterhouse provided. Carter and Ripperger decided they'd keep the bats at the STRI for about two years total and then, with talking sensors in place, release them back into the tree in Tolé from which they were captured.

With an eye to what would happen when they were released, Ripperger and Carter would systematically isolate a bat in captivity at the STRI, starve it for somewhere between twenty-six and twenty-eight hours, and then record which of the other bats, if any, regurgitated a blood meal to its starving group mate. An analysis of the food-sharing social network at the STRI found that, as in Wilkinson's original study, the amount of food a starving bat received from a group mate was predicted by the amount of food the bat had given that individual when it was hungry. In addition, when Ripperger and Carter looked at the grooming network among the bats, they found that the amount of grooming bat 1 had received from bat 2 also predicted the likelihood that bat 1 would feed bat 2 when bat 2 was hungry.

In September 2017, the twenty-three bats from the STRI were brought back to their home tree in Tolé and released. That same day, a control group of twenty-three other bats from that same tree were captured and fitted with talking sensors. As the data flowed from within the roost, a number of patterns began to emerge. For one thing, the STRI-released bats tended to roost together rather than with the control bats. That wasn't especially surprising given that those bats had lived together at the STRI for almost two years: nevertheless, it was reassuring to see that what happened in captivity was relevant to life in the wild. But it was more than simply a matter of which individuals the released bats roosted with. The association patterns among bats were very similar to those seen at the STRI: how much time bats spent near each other in the roost at the tree in Tolé was predicted by the food-sharing and grooming networks the vampire bats had built while they lived in the experimental colony at the STRI.[7]

All of this suggested that the vampire bats were embedded in blood-sharing networks in their roost in the wild. But the blood they were sharing was lapped up far from the roost, and neither Ripperger or Carter or anyone else had any idea whether the network dynamics at the roost affected the vampire bats when they were out at night foraging. The talking sensors made that possible, at least in principle. In practice, there were two related logistical hurdles that had to be overcome before those sensors might shed light on what was happening late at night, out in the pastures, as vampire bats searched for cattle to bleed. Cows tended to roam freely about the pastures in Tolé, and herds per se were a rare occurrence. But what Ripperger and Carter wanted were lots of cows in a localized area, so they had a sense of where the bats would be going to feed. "One of my tasks was hanging out at bars where the locals hang out," says Ripperger, "and I just asked people, 'Do you have cattle?' Everybody either had cattle or has a cousin or a brother who had cattle." Eventually they found someone who agreed to corral a hundred head of cattle so Ripperger and Carter could do their experiment.

Interactions at the roost were localized, but to track the bats from the roost to the area near the cattle required positioning a series of receivers, all powered by car batteries, so that Ripperger and Carter could know where the bats were, should there be a meeting when they were out foraging. But the receivers turned out to be notoriously unreliable: "They would just break all the time," Ripperger says with a smile, "and I was going out every night and checking on these stations every two hours. I felt like I had a baby [out there]." Even when the receivers were working, they seemed jinxed. Once Ripperger saw someone cutting up the antennae and hardware from one receiver with a machete because, as he learned, the machete wielder thought drug traffickers had built it.

When all the glitches were worked out, and data from night meetings away from the roost were analyzed, Ripperger and Carter found that bats with high centrality scores in food-sharing networks at the roost interacted with more bats when they were out foraging. And interactions outside the roost really were "meet-ups," not just bats remaining together as they left the roost. Vampire bats departed the roost individually, not in pairs or groups. And they came back to the roost individually as well. But when they were out foraging, they would meet up for interludes, some short and others lasting up to thirty minutes.

The proximity sensors could only detect distance, not behavior, but there is evidence that meet-ups were associated with clamping onto a cow in a field while a favorite foraging partner does so as well. Audiotapes of a small sample of network members foraging together suggest that meet-ups are initiated by a particular vocalization that vampire bats are known to emit when they are trying to locate preferred partners in the roost. Vampire bat foraging networks appear to extend far beyond the roost: who was meeting up on foraging sorties was in part a function of the potentially lifesaving blood-sharing networks already in place back home in the tree.[8]

While vampire bats suck a bit of blood from domesticated cattle, they don't threaten the economic livelihood of cattle ranchers. Rampaging, networking elephants are another matter altogether.

A few kilometers southeast of Amboseli National Park sit Masai farming communities called shamba. Farmers there grow maize, onions, tomatoes, beans, among other crops. But they have a problem because the park's elephants (*Loxodonta africana*) share the farmers' tastes for those very same crops. And to get those delicacies, some of the elephants raid the farms, which hampers park-community relations since elephants can wipe out a farmer's crops in a few hours. Patrick Chiyo wanted to understand how the raids worked, and to do that he needed to understand the dynamics of the social networks of those marauding giants.

As part of his master's degree at Makerere University in Uganda, Chiyo did a bit of work on crop-raiding elephants at Kibale National Park in Uganda, but he witnessed very few raids during his time there. For his PhD, he decided to study crop raiding from an ethological perspective. He contacted animal behaviorist Susan Alberts at Duke University, who had done work on elephants at the Amboseli National Park in the

Rift Valley Province of Kenya, as part of the Amboseli Elephant Research Project. Alberts was happy to have Chiyo join the cohort of PhD students she was mentoring.

The Amboseli Elephant Research Project began in 1972 under the watchful eyes of ethologists and conservationists Cynthia Moss and Harvey Croze. Because of researchers and tourists frequenting the park, and even more importantly because of the support of the local Masai people, Amboseli elephants were largely spared the decimating effects of the illegal ivory trade that at points reduced the number of elephants in Africa by half. Today Amboseli National Park is one of the few places where elephant calves roam freely, with females nearing their seventieth birthdays and bulls who are in their forties and fifties.[9]

Photographic records of all elephants in the project were available to Chiyo, and the animals were recognizable by tusk size, shape, and natural body marks. Researchers in the Amboseli Elephant Research Project had also collected feces from most of the animals and analyzed the DNA it contained, so that they knew where particular elephants had been, including when they crop raided, as long as they left some feces behind. And they *always* did—in abundance. Exactly the sort of system that would allow Chiyo to dive deeper into the behavioral dynamics of crop raiding.[10]

Elephants have matriarchal societies, and almost all the work of the Amboseli Elephant Research Project had been on females. But from his research at Kibale, Chiyo had hints that crop raiders were more likely males. To find out for sure, and to better understand how crop raiding worked, he recruited local Masai villagers, provided them with cell phones, and had them canvas the local farmlands for information on crop raids. Raids are always at night, and when his assistants learned of a raid, they called Chiyo in the morning. He would then go collect fecal samples. Analysis of those samples showed his intuition was correct: crop raiders were always male.

After he collected all the feces from a given raid, Chiyo would follow the path the raiders took back into the park by tracking the broken branches and plentiful feces the elephants left behind. If he got lucky, he would find them in the park. Soon he began to detect a pattern: if a group of elephants raided a particular farm, they often went back to the same general area in the park, and so that's where he'd go.

Crop-raiding males are a careful bunch. Sometimes Chiyo would be in the park and pick up on signs that a group of males was off to raid. "Unidirectional movement, poop, poop, poop" is how he describes it. Chiyo followed and, on occasion, he'd catch up with the putative raiders when they neared a shamba, where "they'd stand and freeze and wait for darkness." For good reason too. Even though raids are under cover of darkness, they are still dangerous. Masai farmers are understandably terrified that their main source of subsistence can be wiped out in a night and they guard their farms as best they can. A notorious crop raider that Chiyo had dubbed Adam knew that better than most: the wounds on Adam's flank were from spears courtesy of a vigilant farmer who saw a raid unfolding.

Between June and December of 2005–2007, Chiyo and his colleagues collected data on the composition of crop-raiding groups, from both direct observation of raids and identification of individuals based on DNA analysis of feces. During that same period, Chiyo would often spend a few months at Duke, where he learned all about social networks. To build the elephant social networks, he and his colleagues used the association patterns (outside the period of raids) of twenty-one known elephant marauders and thirty-seven elephants who never raided. When they measured the density of the social network—the number of observed pairwise associations divided by the number of all possible associations—Chiyo and his team found that the network of all fifty-eight males had weak ties and so was only loosely connected. But probing down one level, they discovered there were six cliques embedded in the larger network. And crop-raiding behavior, which provided raiders with huge bounties, was to a large extent what structured these cliques: each clique had at least one known crop raider, and three of the cliques had about the same number of crop raiders as those who did not raid.

Network structure in cliques was especially valuable in predicting who would become a raider. The probability of being a raider was significantly higher for elephants whose favorite partner in a clique was a raider than when it was not, and similarly for males whose friends of friends were raiders. But there was more to it than that. Crop raiders were, on average, older than non-raiders, and the odds of becoming a raider increased with age, particularly when males reached the age of sexual maturity, likely because of the extra need to compensate for the energetic costs of finding a mate. What Chiyo and his colleagues discovered was that the chances of a young male becoming a raider increased when its close associates in a clique were older raiders, suggesting that young crop raiders in training were most often learning the craft from older, more experienced individuals. Given the spear marks on Adam's flank, this seems prudent.[11]

As humans seemingly lay claim to every square kilometer on the planet, interactions with resident animals are on the rise, and not only between elephants and farmers on shamba. These interactions rarely work to the benefit of the inhabitants present before our colonization. But, on occasion, humans may improve one or another aspects of animal life. Bird feeders in suburban and urban settings, for example, provide individuals in many species with more food than their counterparts can procure in rural settings. That's certainly the case for the black-capped chickadees (*Poecile atricapillus*) that animal behaviorist Julie Morand-Ferron studied in both rural and urban grounds around Ottawa (Morand-Ferron died far too young at age forty-four in 2022.)

Black-capped chickadees are songbirds, and when Morand-Ferron arrived at the University of Ottawa in 2012, she was already versed in working with this group. As a postdoctoral fellow at Oxford University, she had done work with another songbird, the great tit (*Parus major*), in the Wytham Woods near the university. Morand-Ferron's main research interest when she was at Wytham was the cognitive abilities of the great tits, but she also teamed up with Lucy Aplin and Ben Sheldon to study the social networks of these birds. "Naively," Morand-Ferron told me in 2021, "I thought that what I had learned on the great tits could be transferred to the black-capped chickadee." It couldn't. The great tits moved around a lot, even in the winter, living in what ethologists call a "fission-fusion" society. Not so for the winter flocks of black-capped chickadees in Ottawa, who, once the first cold snap hits in October, stay put in closed groups.[12]

After reading Susan Smith's 1991 book on black-capped chickadees and doing a bit of basic work herself on their behavior, Morand-Ferron realized that across the span of just a few kilometers, chickadees in urban parks were leading the good life, at least in terms of foraging, compared to the birds in nearby rural settings. In urban settings, food was available all the time, and there was more of it than in rural settings. Soon Morand-Ferron was studying chickadees at four parks near downtown Ottawa—all abutting residential areas full of bird feeders—as well as at four rural, forested sites that were at least 16 kilometers from downtown. Each fall she'd capture the chickadees using mist nets, fit each with a passive integrated transponder (PIT) tag, which activates when near any of the radio-frequency identification (RFID) receivers Morand-Ferron had placed on feeders that had not been mangled beyond recognition by hungry squirrels. Then, she would periodically refill the feeders and download the data on when chickadees visited.[13]

It wasn't long before Morand-Ferron was seeing things that made her think that the social behavior of chickadees at rural and urban sites was different. When she and her team were out catching chickadees to implant PIT tags, they noticed that birds at urban sites were arriving all the time, but when they were at rural sites, they'd often find no chickadees, and then suddenly a large flock would visit. What's more, the flocks at the rural site seemed to act more like a cohesive whole than their counterparts in urban settings. When Morand-Ferron drove up to the rural sites, always in the same lab car, the chickadees would swarm around a feeder and "they'd literally ask for us to fill the feeder." Not so in the parks near the city. "We thought, yes, that makes sense . . . if you live in town and you're near the suburbs, there are plenty of feeders. You don't need to rely on social information. . . . If you live in the woods, then being attentive to social information is more important for your survival."

All of this got Morand-Ferron thinking about the social networks she had dabbled with at Wytham. What she had with the black-capped chickadees were stable flocks of birds, loyal to a general locale, in one of two very different types of environments, presenting her with the opportunity to see how environment affected the dynamics of the feeding networks the birds formed. And there was a theory that generated a testable prediction: when food is less reliable and less abundant, as in rural sites, getting information from others in a foraging network will be especially important.[14]

Morand-Ferron and her colleagues set out to test that hypothesis. Under the cloak of darkness, so birds didn't notice, they moved the feeder they had placed on a territory to a new location about 100 meters away. Using data from more than 80,000 visits of seventy-three chickadees, they examined how quickly information about the new location of food spread through a network. Birds in both environments found the relocated feeder, but the *rate* of information transfer about the new location was higher in chickadee networks at rural sites than urban ones: as theory predicted, when there's not much food and it is hard to know where to find it, social networking is especially important.[15]

For a more unusual, indeed unique, way that humans interact with nonhuman social networks, we return to the bottlenose dolphins we met at the start of the chapter, but move from New Zealand to Brazil. There, for more than three decades, ethologist Paulo Simões-Lopes has been studying dolphin populations in the lagoon systems along the coastline near Laguna, about 800 kilometers south of São Paulo. The dolphins in nine populations along that stretch do something that no other dolphins—and almost no other animals, period—do. They not only network with each other, but cooperate with humans to secure more food for both themselves and their primate partner.

Each autumn, a huge mullet migration takes place in southern Brazil. Both the dolphins and the fishermen see the fish as prize prey. Up to fifty fishermen, wading waist deep in very cold water, wait for the chance to cast large circular nylon nets called tarrafa over schools of mullet. The problem for the fishermen is that the water is murky, and it is next to impossible to see the fish. The problem for the sixty or so dolphins at Laguna is that compared to their other prey, mullet are large and hard to catch. But dolphins aren't especially troubled by murky water, as they detect mullet using echolocation, a built-in sonar system that would be the envy of most engineers.

Dolphins produce sound waves in their nasal sacs and focus those waves through fatty tissue and fluid in their foreheads. Once the sound waves are shot out into the water, they travel until they bump into an object, at which point they bounce back to the dolphins, who use their lower jaw as a receiver. From the lower jaw, the waves travel to the inner ear and then to the brain. Objects of different sizes and densities reflect back sound waves of different frequencies, and the dolphins use that information to "see" what is in the water around them. When their sonar detects mullet, dolphins signal fishermen that the fish are present by curving their backs and then slapping their heads or their tails on the water surface. The fishermen then cast their tarrafa and pull in loads of mullet. The confused mullet who escape the tarrafa often swim right into the mouths of waiting dolphins. It's the perfect win-win situation.[16]

Laguna newspapers from the late 1890s featured articles about this dolphin-human mutualism, and so Simões-Lopes knows that, at the very least, it has been going on for more than 130 years. And though many dolphins don't signal fishermen, every fisherman knows which dolphins do. "It is famous [in southern Brazil]," Simões-Lopes says. "I grew up watching those dolphins. . . . I would sit on a rock in the canal and watch for hours. I knew it was unusual. . . . I knew there were dolphins in a big harbor farther south where dolphins and fishermen don't interact."

Today Simões-Lopes has a team of ten working with him,

but he began on his own in 1988. Soon thereafter, he entered a PhD program and built his dissertation around his research on the dolphin-human foraging mutualism. Each day he brought a folding chair with him and set it up on a rock, watching the dolphins through his binoculars, taking photos—he had compiled a mug book with photos of all the dolphins in the lagoon—and filling notebook after notebook with data on dolphins signaling fishermen.

Simões-Lopes began to know the fishermen, and they began to know him. He also was starting to get a good feel for which dolphins at Laguna signaled the fishermen and which did not. Not surprisingly, the fishermen also kept tabs, telling Simões-Lopes about the "good dolphins" (who signaled fishermen) and the "bad dolphins" (who did not). The fishermen know not only which dolphins signal, but which dolphin will give which signal: "Each dolphin gives the signal in a different way," one fisherman said, "and we need to know [the different signals] in order to catch the fish." Another fisherman was more of a romantic, telling Simões-Lopes and his colleagues, "This is beautiful. It doesn't happen everywhere."[17]

The more that Simões-Lopes thought about those "good" dolphins and "bad" dolphins, the more he wanted to understand them better. Years later Mauricio Cantor joined Simões-Lopes's team; Cantor had worked with Hal Whitehead, a leader in early social network analysis. Simões-Lopes and Cantor decided that a network analysis might help them delve deeper into the between-species cooperation they observed on a daily basis. In 2008 they contacted David Lusseau, who had done the network studies on bottlenose dolphins in New Zealand, and asked if he would be interested in acting as a sort of conceptual consultant specializing in social networks. Lusseau was more than happy to join their team.[18]

Simões-Lopes and his team assumed dolphins learn how

to signal humans from other signalers they associate with, so for their social network analysis, they were especially interested in whether signaling dolphins preferred spending time with other signaling dolphins, both when they were chasing mullet into nets and, just as importantly, when they were not. To test whether there were cliques of signalers and cliques of dolphins who didn't signal, Simões-Lopes's team looked at clustering coefficients of sixteen cooperators and nineteen dolphins who did not signal and cooperate with fishermen.

What they discovered were three cliques within the larger network of the thirty-five dolphins. Clique 1 had fifteen dolphins: each and every one of them cooperated with the local fishermen. Dolphins in this clique associated with one another not just during the autumn mullet fishing season but the rest of the year as well. Clique 2 had a dozen dolphins, *none* of whom cooperated with fishermen, and dolphins in this clique were not as well connected to one another as the individuals were in Clique 1. Clique 3 was made up of eight dolphins: seven never cooperated with fishermen, but one—dolphin 20—did. And of all thirty-five dolphins in the network, it was dolphin 20 who spent the most time interacting *across* cliques, acting as what Simões-Lopes and his colleagues call a "social broker" between the signalers and non-signalers.[19]

This behavior is all wonderfully complex, and we humans—and I don't just mean the artisanal fishermen of Laguna—should be grateful to play a role in understanding it.

Social networks affect foraging in everything from elephants and chickadees to vampire bats and dolphins. From an evolutionary perspective, one reason that strategies to procure lots of food are favored by natural selection is that they provide animals with the energy to find a mate, reproduce, and, in species with parental care, rear offspring. And so, it is not surprising that mating, parental care, and more are also influenced by the structure of the social network an animal is embedded in: indeed, animals actively manipulate networks to increase their reproductive success. In the next chapter, we'll start off seeing how they do this by peering in on some kangaroos at the Great Sandy National Park in Australia.

4. The Reproduction Network

The true character of a society is revealed in how it treats its children. —Nelson Mandela, address at Worchester Station, September 27, 1997

It's a long way from the University of Michigan in Ann Arbor, where Anne Goldizen did her PhD work, to Queensland, Australia. But Goldizen found herself there on a postdoctoral fellowship, where she was studying the mating system of Tasmanian native hens (*Gallinula mortierii*). The birds were fascinating and Tasmania itself was a wonder, except for one thing: there was almost no one else in Tasmania who was studying animal behavior, and that just wouldn't do.[1]

To remedy the intellectual loneliness, Goldizen reached out to ethologist and conservation biologist Peter Jarman, about 500 kilometers northwest of Sydney, who happily took on the role of friend and adviser. On one of Goldizen's visits to Jarman, he set it up so that she could spend a few days with his field assistant at a research site near New South Wales's Wallaby Creek, where he was studying the ecology and behavior of eastern grey kangaroos (*Macropus giganteus*). "Peter's field assistant would tell us this is so-and-so, and who [so-and-so] interacted with," she recalls fondly. "It was fantastic." Then it was back to studying native hens in Tasmania, but the kangaroos were never far from Goldizen's thoughts.

In time, Goldizen decided that she couldn't bear to leave Australia and decided to settle down there, landing a faculty position in Queensland. Jarman's field site was an eight-hour drive from there, but just 160 kilometers north from her academic home was a population of a hundred eastern grey kangaroos waiting to be studied at Elanda Point, near Lake Cootharaba, in an area surrounded by Great Sandy National Park. Together with her student Alecia Carter and others, Goldizen began work on the social behavior of females, with thoughts of digging into social networks at Elanda Point.

Female eastern greys defend territories, but they also forage in small groups, where membership is very fluid. Goldizen's team found that females had favorite foraging partners, but before they could learn much more about those associations, the dingoes came. And they were very hungry. During an especially bad drought, the starving dingoes ramped up their hunting at nearby Great Sandy National Park and decimated the kangaroo population Goldizen was studying, killing a large portion of the adult females. Though Goldizen was able to salvage something interesting from that unfortunate series of events—females were much less selective about their foraging partners after the dingoes wreaked their havoc—she'd have to go elsewhere if she wanted to continue her dive into eastern grey kangaroos and their social networks, including how those networks affected parental care. She did just that, moving her operation to Sundown National Park at the Queensland/New South Wales border, a four-hour drive southwest of her university. But before we join Goldizen and her team in Sundown National Park, let's look at a few other reproduction networks.[2]

Katavi National Park in Tanzania isn't where ethologist Miho Saito first encountered giraffes. Saito's fascination with giraffes started much earlier. When she was three years old, her father's work landed the family in Nairobi, Kenya, and young Miho often visited a place where little children could feed giraffes by hand. When Saito's family returned to Japan, she would look at photos from her time in Nairobi and dream of one day returning to Kenya to see wild giraffes. In 2010 she got her chance when she entered a master's program at Kyoto University as a student of Gen'ichi Idani. Idani was well known at the time for his work on chimpanzees and bonobos, and he operated a base camp in Ugalla, Tanzania. As fascinated as Saito was with chimpanzees and bonobos, it was giraffes that were calling her back to Africa, and Idani was happy to have her follow her passion. Saito joined Idani when he went to Tanzania in 2010. After he taught her the lay of the land, she was on her own for the next five months with her study population of giraffes at Katavi National Park, about 300 kilometers south of Idani's base camp in Ugalla.

Katavi, the country's third largest national park, sits in southwest Tanzania. The woodland forests there, called miombo, are brimming with plants from the genera *Grewia*, *Markhamia*, and *Combretum*. Cape buffalo, lions, elephants, spotted hyenas, wild dogs, leopards, eland, zebras, impalas, reedbuck, and, of course, giraffes (*Giraffa camelopardalis tippelskirchi*) roam Katavi's 4,500 square kilometers, while African fish eagles, lilac-breasted rollers, emerald-spotted wood doves, black-bellied bustards, and greater painted-snipes soar above.[3]

At Katavi, Saito rented a room in the home of a local family who lived in the nearby village of Sitalike. Each morning she would walk ten minutes to the main office of the park, where she would meet with a ranger. Because of the danger presented by lions, elephants, and water buffalo, an armed ranger would always accompany her while she spent seven hours a day, divided into a morning and afternoon shift, working at the park. A day with Saito meant a lot of walking—15 kilometers or more—and some of the rangers preferred their more standard routine, in which they patrolled about the park in a car. Other rangers made the best of it. One who had quit junior high school, but reconsidered the errors of his youth and was studying to take the high school entrance exam, "just brought his textbooks and notes," Saito says, "and while I was taking giraffe data, he was studying."

Early on, Saito just needed to get to know one giraffe from another. Adults can be recognized by their unique fur patterns, and juveniles can be aged by the size and shape of their ossicones, the two skin-covered bones projecting from the tops of their skulls. Soon Saito gathered size data and drew sketches of about 150 giraffes and kept tabs on a subset of 70.

Adult male giraffes spend a lot of time alone, but adult females are often found in small groups of about five. When Saito would spot a group of female giraffes, she'd watch them through her binoculars. Any giraffes within 200 meters of one another were considered part of the same group, and Saito would take notes, with pen and paper, on everything group members did, including the behaviors they displayed.

From her time at Katavi in 2010, and then again from a similar stint in 2011, Saito came to learn that female groups were very fluid, with individuals often moving from one group to another, and that females were not especially loyal to place. But there was an exception to both rules. In what have been dubbed nursery groups, a group of lactating females, each with a single calf, stay put in the same location, day after day, for about three months. Saito found that life in nurseries provide added protection for calves above and beyond that provided by their mother: when a calf's mother would leave for an hour to feed or drink, another lactating female would step up and act as guardian over the calf, being especially vigilant when a nursery was in an area of the park associated with lions, the giraffe's main predator. Another benefit of calf life in a nursery is that although females show a very strong tendency to allow only their calf to suckle, from time to time calves receive milk from a female other than their mother.[4]

Saito could see there were differences in the social behaviors of nursing females and other females, but she wasn't sure how to dig deeper into the long-term effects of nurseries on giraffe social dynamics. That changed when she began her PhD work with animal behaviorist Fred Bercovitch as one of her mentors. Bercovitch not only had experience working with giraffes, but also was familiar with social network theory. Soon Saito was reading the social network literature and began thinking that network analysis might be a powerful tool for understanding her giraffes. When she returned to Katavi for her PhD work in 2016, 2017, and 2019, she did so with a social network mindset.

Saito knew that for the three or so months that females and their offspring were in a nursery, they interacted often and formed well-defined clusters within the larger population at Katavi. But what about when nursery groups disbanded? Young giraffes don't wean until they are around a year old, and they tend to stay with their mother until about eighteen months old, so bonds between offspring and mother would be strong for at least that long. But what of the interactions between adult females who had recently been in a nursery together? Did they tend to group together after their nursery days were over? Did the effects of nursery networks outlive the nurseries themselves?

Saito looked at data from one nursery per year for each of the field seasons she had worked at Katavi, and she calculated the average strength of connections between females from the same nursery after that nursery had disbanded, but when juveniles were still spending virtually all their time with their mothers. She looked at the same network measure in the *same* female after her offspring had gone off and joined a group of his or her peers at about eighteen months of age. Compared to females who had not recently been in a nursery, females who had shared time together in a nursery had significantly greater weighted tie values to each another in the overall giraffe network when their calves remained with them, but not when their little ones, as it were, had gone their own ways: networking in nurseries affected moms after nurseries broke up, but only for so long.[5]

Networks associated with reproduction aren't only about parental care. In the house finches (*Haemorhous mexicanus*) that behavioral ecologist Kevin Oh studies, social networks are also about selecting a mate.

House finches are a scrappy lot. In the late 1930s, their range was expanded to the East Coast of the United States, initially by no will of the birds, when a bird dealer in California illegally shipped some house finches to the Long Island area of New York. Then the birds took over. The first house finch sighted living wild was spotted at Jones Beach on April 11, 1941: a year later, seven house finches were seen in Babylon, about 25 kilometers away, and soon a house finch nest was discovered farther away still. The birds on Long Island had a darker, duskier look to their plumage than those from

California, but eventually, when someone thought to capture and wash them, it was discovered that the duskier coloration was soot: when the birds were cleaned up, they were dead ringers for those in California. By 1951 there were almost 300 house finches living in four separate areas on Long Island. Today the Cornell Lab of Ornithology lists about 17,000 house finch sightings *per week* in the Long Island area, and almost 400,000 sightings per week for the state of New York, all of which is to say this is a bird that knows how to make the best of it.[6]

As a kid, Kevin Oh recalls getting a small microscope as a Christmas present one year. This little device had the added feature of an adapter that turned it into a crude projector. Looking at blurred images of water fleas, plant roots, insects, and more projected onto his bedroom wall turned Oh into an aspiring scientist. A decade later, he was a premed major at Bowdoin College in Maine and looking for some research experience. As luck would have it, he took an ornithology class, taught by Nathaniel Thoreau Wheelwright, that really sparked his interest.

Wheelwright and others ran the Bowdoin Scientific Station on Kent Island in the Bay of Fundy, across the Canadian border, about a half day's drive from the college. Soon Oh and "a gaggle of undergrads," as he describes his fellow students, "would spend the summer doing projects and interacting with visiting scientists and postdocs." He did a bit of work on territoriality in savannah sparrows (*Passerculus sandwichensis*) and quickly realized that fieldwork in animal behavior allowed him to marry his love for science with his passion for the outdoors.

When Oh began his PhD studies at the University of Arizona in 2003, his mentor, Alexander Badyaev, had joined the faculty only a year earlier. Badyaev had done his own PhD work on house finches, and when he got to Tucson, he decided to start up a study on the thousands of house finches on the campus of his new academic home. The first step was to band all the finches so that each one could be recognized. Badyaev told Oh that if he would take the lead on that, it might help him develop ideas for a dissertation. At first, Oh wasn't sure, as he knew many fellow students who were heading to exotic places around the world to do their research. But soon he was on board, trapping and banding house finches by placing one aluminum and three plastic color bands on one of the bird's legs. By the time he finished his PhD six years later, Oh and others had banded more than 10,000 birds.

Getting a large-scale field study off the ground requires more than just banding birds. Feeding stations were placed in a grid around campus. And nest boxes—a lot of nest boxes—were built and put out. The finches bred in shrubs and trees around the campus, but they loved the nest boxes. Once enough finches were banded, and enough feeders and nest boxes were in place, Badyaev's team, led by Oh, gathered basic morphological measurements and took blood samples for DNA analysis on a subset of the birds.

After he got to know the birds, Oh's ideas for a dissertation project began to crystallize around social networks, mate choice, and reproduction in the finches. Prior work by ethologist Geoff Hill had found that colorful male house finches were more likely to obtain a mate than their drabber counterparts, but Oh wanted to know if there was some way that drabber, less attractive males might game the birds' social network system to increase their relative attractiveness. It was a daunting task that involved not only measuring male coloration, but tracking one-year-old males when they flocked and roosted in mixed-sex groups, so as to get data on networks outside the breeding season, and then finding, or trying to find, every nest on campus during the breeding season to estimate male reproductive success.[7]

The network data on mixed-sex flocks outside the breeding season was gathered at the feeding stations. Oh and his colleagues placed "walk-in traps" about a cubic meter in size near the stations. House finches would enter through a tunnel and then get trapped inside. Oh would then take the trapped birds out and place them in a paper bag. Next, they'd place the cage filled with bags on a little kiddie trailer attached to the bikes they used when navigating campus and bring them back to the lab. "Picture this chicken wire cube," Oh says, "with twenty to thirty little paper bags hopping up and down in the bottom of it" as they pedaled back to the lab to weigh the birds, take photos, and collect other data. When all that was done, the birds went back into the bags and onto the kiddie trailer again for a short ride back to freedom at the site where they had been captured.

Oh and Badyaev learned that during the nonbreeding season, the house finch social network across campus was made up of about two dozen groups, with an average group size of about thirty birds. Many, but not all, behavioral interactions were between finches from the same groups. Females tended not to interact with birds in other groups, displaying what Oh and Badyaev call low social lability. For males, social lability depended on plumage color: drabber males were more socially labile than colorful males. But why? Oh and Badyaev mulled over the idea that less colorful males may have been forced out of groups by more colorful males. They weren't: less colorful males tended to be behaviorally dominant to more colorful males, and so were not being bullied and forced to interact with those outside their group. Instead, social lability was an active choice on the part of less colorful males. With that piece of the puzzle in place, Oh could now tie his interest in mating and reproduction to what was known about network dynamics. To do that, he and Badyaev focused on networks in January and February, when pair formation occurs.

Females breeding for the first time form pair bonds with males in their own group as well as with socially labile males from other groups who happen to be around when they're assessing potential mates. All things being equal, females prefer males with brighter plumage, but all things are not equal because drabber males are more socially labile. What that translates into is that lability had no effect on reproductive success in colorful males: they did just as well if they stayed put or moved about interacting across groups. But for drabber males, lability was important. The more labile a drab male was, the higher his reproductive success, likely because less colorful males were sampling groups to find one where their *relative* attractiveness was highest during the period when females were most actively selecting mates.

Oh and Badyaev plotted male reproductive success on a three-dimensional graph that looks like a topographical map with longitude, latitude, and altitude. When social lability is plotted on one axis of the graph, plumage coloration on another, and reproductive success on the third, there are two peaks that jump off the map: one of those peaks is associated with high reproductive success in drab, labile males and the other peak with high reproductive success in colorful, less labile males.

Colorful males stay put and reap reproductive rewards, but drabber males need to be more proactive, and they do that by rewiring the network, using social lability to manipulate network structure to construct their own reproductive peak.[8]

Kangaroos at Sundown Park in Australia aren't rigging the system, but Anne Goldizen set out to find what was going on in their social networks. Even surrounded by the natural beauty of the kurrajong trees and spotted hyacinth, with bowerbirds, striped honeyeaters, red-winged parrots, and superb lyrebirds above, it was not easy going, particularly in the summer, when temperatures routinely hit 40°C. It didn't help matters that the ranger's house where Goldizen's students Emily Best and Clementine Menz stayed had no air-conditioning or, for that matter, much of anything else—except mice. Loneliness was also a problem. Though a national park, Sundown had very few visitors, and almost none came near the area of the park where Best, and later Menz, worked. Goldizen insisted that when either of her students went to spend time at the park, they needed to have someone accompany them, both for safety reasons—there were lots of deadly snakes slithering around—and to provide companionship.

During the summer, kangaroos didn't like the scorching temperatures any better than Best did, and they were active for about two hours in the morning and another two hours in the later afternoon. And active for kangaroos means feeding. Fortunately, where they fed much of the time was a stone's throw from the ranger's cabin. Day in and day out, two weeks out of every month for two years, Best gathered data on which kangaroos fed together.

While Best, and later Menz, were gathering data on social structure, Goldizen was thinking about how to analyze it. She started probing the social network literature and quickly realized much of it focused on primates, birds, and cetaceans. But she knew that the kangaroos she and her team were studying lived in a complex society and thought it likely that networks structure that society.

As with the kangaroos at Elanda Point, female kangaroos at Sundown National Park defended territories and foraged in small groups, and group composition could change from hour to hour, even minute to minute. And again, like female kangaroos at Elanda Point, some females had favorite foraging partners, which Goldizen and her team quantified using a social network measure called the half-weight index. But the foraging network dynamics were more complicated at Sundown than at Elanda Point. When clustering coefficients were calculated to assess possible cliques within the larger network, females who had strong partner preferences tended to be found together. Clearly, a female would often be found near *her* favorite partners. That need not translate to her also associating with other females who had stronger partner preferences. But it did. Females with strong partner preferences liked other females with the same proclivity toward forging close friendships.

The more Goldizen and her team understood about the kangaroos' foraging networks, the more they wanted to understand the consequences of networked foraging. What they found surprised them, in part because of how decisions made about foraging buddies affects parental care and survival of joeys. Contrary to what they predicted, the number of favorite friends a female had was *negatively* correlated with the probably that the female's joey survived to adulthood. In particular, the strongest negative effect on offspring survival was between when a joey was too large to stay in its mother's pouch and the time it weaned completely from its mother's milk.[9]

Goldizen thinks two things explain how preferences displayed in foraging networks influence joey survival. Sociable females, with lots of preferred foraging partners, may be more likely to produce more sociable offspring. If, because of their proclivity for interacting with others, those offspring spend more time away from their mother before weaning, that could lead to a greater chance of being killed by predators.

But networking with favored partners also takes time and energy that might be better used for parental care. Mothers who have strong preferences for certain foraging partners may be more likely to lose track of joeys while they search for their preferred partners. That not only reduces the amount of time that a joey and its mother spend together, translating into less milk for young kangaroos, but it almost certainly makes joeys easier targets for predators.[10]

From giraffe nurseries to kangaroo friendships to male finches manipulating the competition to secure more mates, social networks affect so many aspects of reproduction in animal societies. At the same time, where there is life in societies, there are struggles for power. The effects of power struggles ripple through societies, and social network analysis is the perfect tool for examining how and why, and what it all means.

As a case in point, we'll begin by moving from Australia over to New Zealand and sneak a look at what's happening in the societies of the Australasian swamphen, also known as the pūkeko.

5. The Power Network

The Party seeks power entirely for its own sake. We are not interested in the good of others; we are interested solely in power, pure power. . . . We know that no one ever seizes power with the intention of relinquishing it. —George Orwell, *Nineteen Eighty-Four* (1949)

Māori folklore has much to say about the strikingly colorful pūkeko (*Porphyrio melanotus*), a bird that calls New Zealand home. Also known as the Australasian swamphen, the pūkeko is said to have gotten its deep red beak color when a high chief named Tāwhaki lay dying and his blood marked the bird's bill. Another legend, "How the Kiwi Lost Its Wings," tells the tale of the pūkeko's preference for swampy habitats. Tänemahuta, god of the forests, was angered that his children, the trees, were being ravaged by bugs. He asked many different types of birds to come down from the trees and live on the ground to rid the forest of the bugs. When Tänemahuta approached a pūkeko, the bird looked at up at the sun, then down at the forest floor, and told the forest god, "No, Tänemahuta, it is too damp, and I don't want to get my feet wet." Tänemahuta continued his search until, at last, the kiwi agreed to help him. When the forest god doled out punishments to those who had denied him, the sentence for the pūkeko matched the crime:

"Because you didn't want to get your feet wet, you will live forever in the swamp."

For a swamp lover, there is much to be said for Tāwharanui Regional Park on New Zealand's North Island. About an hour's drive north of Auckland, and just 3 kilometers from Kawau Island, the park is bordered by the Pacific Ocean to the east, and the Tāwharanui Peninsula to the west. Pūkekos love the wetlands found throughout the park and happily form their group territories there. Animal behaviorist Cody Dey was just as happy to study the complex social networks they form on those territories.

For his PhD work, Dey was interested in cooperative breeding, in which "helpers" assist in raising young who are not their own, often delaying their own chances of reproducing in doing so. Pūkekos were perfect for looking at questions about cooperative breeding, and not just because of the allure of fieldwork on the other side of the planet. It was what the pūkekos were doing that was the real draw. Not only are pūkekos cooperative breeders, providing care to offspring who are not their own, but they are also "joint layers," the rarest type of cooperative breeder, in which multiple females lay their eggs in the same nest.[1]

Dey's first stint working with pūkekos at Tāwharanui Regional Park was for six months. Though Tāwharanui is a public park, Dey was rarely disturbed by passersby, in part because most people were at the park to surf, and the pūkekos mainly hung around more inland, near fenced-off paddocks where cattle grazed. For those New Zealanders in the park who were there as bird lovers, pūkekos were an afterthought. "They're kind of like the Canada geese of New Zealand," Dey says. "People just want to burn past them to see kiwis."[2]

Groups of pūkekos, including joint-laying females, defend territories that often abut the cattle paddocks in the park.

They vigorously defend those territories: sometimes it's just the matter of a pūkeko standing at its territory posturing and squawking at birds in an adjacent group, but other times full-scale battles break out. It's a more peaceful time for a pūkeko when birds from different territories feed in the same paddock, where they show none of the aggression they do when defending home.

Every morning at sunrise, Dey would set up a temporary observation blind, ideally on a hilltop where he could look down on a paddock and the surrounding area. Other days he would set the blind just on the other side of a paddock fence. No matter where he set up the blind for a day, Dey would peer through a spotting scope and gather data on power, courtship, nesting, and other behaviors, both when birds were in their territories and when they were on neutral feeding grounds.

It wasn't long before Dey realized that he had come to New Zealand with two misconceptions about pūkekos. For one, he had thought groups would be relatively small, on the order of four or five individuals: some were, but many were larger, with a dozen or more birds sharing a territory. Also, because of his interest in joint laying, he expected birds living on the same territory would generally be nice to one another, but he soon learned that the quest for power was key and it often manifests itself in rather nasty behaviors. The dominant of two pūkekos would peck, kick, and charge, while the subordinate would display submissive behaviors, including both retreating from the dominant pūkeko and changing its posture so that its head and bill pointed downward, rather than upward as in dominant birds. Dey also came to understand that for males, the red fleshy "frontal shield" that sat above their beak was a signal, and that larger shields indicated not just status (as well as larger testes size), but a higher slot in the power structure.[3]

Dey became interested in what the relationship was between joint laying and aggression. When he followed chicks hatched in a nest that had multiple females contributing eggs, what he found was that having a powerful mother mattered: chicks of dominant females hatched earliest and grew faster, had higher survival rates, and were more likely to become dominant later in life than chicks of subordinate females. Hatching order and its effect on subsequent dominance provided some insight into dominance in pūkekos, but for Dey it wasn't enough, because it didn't capture the dynamics of power at the group level. Fortunately, another of Dey's interests—social network analysis—came to the rescue.

Dey tinkered with some social network software so that it could better analyze power dynamics in his pūkeko networks. Rather than merely figuring out who was the dominant and who was the subordinate in every pair within a group, he tweaked the software so that it would look at three pūkekos at a time and test whether dominance interactions in a troika was "transitive." For three individuals—1, 2, and 3—a power hierarchy is transitive when if 1 is dominant to 2, and 2 is dominant to 3, then, closing the power triangle, 1 is dominant to 3. What makes these sorts of transitive threesomes especially interesting is that when large groups are composed of transitive threesomes, then the overall group will have a very linear power structure, where 1 dominates 2, who dominates 3, who dominates 4, all the way to the lowest-ranking member of the power hierarchy, who, alas, dominates no one.

Dey collected data on aggression in eleven different pūkeko groups that ranged in size from four (where there are only four triads) to thirteen (where there are many, many different triads). To facilitate aggressive interactions, just before he gathered his data, Dey tossed some corn on the territory of the birds and then commenced to noting aggressive and submissive behaviors displayed by group members.

In both years of his social network study, troikas were most often transitive, leading to well-defined power structures (linear hierarchies) in all eleven groups. While there is debate among ethologists whether such large, stable hierarchies reduce overall aggression in a group enough to benefit all group members, one thing is for certain: for pūkekos, the dynamics of power are rooted, at least in part, in social networks.[4]

Before we look at other ways in which networks are important in power dynamics, we should have a more explicit definition of power. By power, I mean the ability to direct, control, or influence the behavior of others and/or the ability to control access to resources. The subtle and not-so-subtle ways that animals work to garner power are astonishing and informative because they provide an evolutionary window through which we can better understand behavioral dynamics in group-living species. Social network analysis provides animal behaviorists with a tool to understand the complexities of power: a tool that has been put to good use when studying field cricket power struggles.[5]

Should you find yourself rambling around a meadow in Spain and happen to trip on one of what you come to discover are, in fact, 120 low-to-the-ground, closed-circuit television cameras pointed squarely at cricket borrows, you'd only be half wrong if you assumed someone was filming a reality television show about life in an insect neighborhood. Reality shows revolve around social networks, and if you want to study cricket social networks, which David Fisher and his colleagues Rolando

Rodríguez-Muñoz and Tom Tregenza most certainly do, you don your directors' hats and roll the cameras.

Those cameras are there to spy on the everyday minutiae of life for a few hundred field crickets (*Gryllus campestris*) who live in that meadow on the northwest slope of a valley in the Asturias region of northern Spain. In 2005 the first set of cameras were placed in the meadow, each pointing to at least one burrow. Every burrow has a tiny flag with a three-digit number adjacent to it, so the researchers know which burrow is which. They also know the identity of every cricket, as each is briefly captured, a DNA sample is taken, and a unique very tiny alphanumeric tag is affixed to its back before the cricket is released back into its burrow. Cameras record all interactions in an area of about 60 square centimeters around a burrow, and then, via an intricate set of fiber-optic cables laid out in the field, all cameras feed video to a bank of computers in a small house nearby. During the mating season, which runs from late April to early July, tens of thousands of hours of interactions around burrows are recorded.[6]

Field crickets live for a year, and they produce a single clutch of eggs. The immature nymphs spend the winter in a burrow, completing metamorphosis in the spring. As sexually mature adults, males leave their burrows during the day and call to attract a mate. They also move about the meadow searching for a burrow that might have a lone female with whom they can mate. Males are also always on the lookout for ways to co-opt a better burrow—one that's deeper, closer to a food source, and, most importantly, has a female residing in it, even if there happens to be a male already in there with her. Not surprisingly, male residents of such burrows are not keen on giving up their home or their mate, and power struggles—with flaring mandibles, charges, and grappling combatants—are common. Fortunately, male contests often occur aboveground near the burrow entrance, so the spy cameras can catch all the action.

David Fisher joined the cricket project in 2012. He was searching for a place to start his dissertation work when he came upon an advertisement that animal behaviorist Tom Tregenza at the University of Exeter had placed. Tregenza was looking for a PhD student to join him and Rolando Rodríguez-Muñoz on their project and, more specifically, to look at social networks in the crickets. The more Fisher looked into the cricket system, the more intrigued he became, as he began to realize that Tregenza and Rodríguez-Muñoz "basically take all the things we typically do with crickets and similar model systems in the lab and do it in the wild." Soon Fisher was off to study at Exeter, where he read up on the social network literature, and then to Spain, where he was tagging crickets and watching videotapes. A lot of videotapes. He kept a special eye out for fights and matings, because not only do power struggles happen aboveground due to cramped space in burrows, but all matings also occur aboveground near the burrow entrance.

Outside of the cricket mating season, Fisher was glued to a bank of screens, transcribing the events from the seemingly endless videos that had been shot the prior mating season. Some days he snuck in another of his loves that, by chance, focused on a different sort of cricket: "England was playing cricket in India, so I would get up at like five or so in the morning and listen [to the games]." When that was done, it was time to turn to the monitors. "A cricket would leave its burrow and [I'd] record it as 'A1 left burrow 32 at 10:05,'" Fisher describes the process, "and then, an hour later, I'd be like 'cricket A1 arrived at burrow 58 at 11:30.' . . . Because they're moving around all the time, we can then build up these networks . . . male fighting networks . . . male/female mating networks." The hours of tedium were sometimes broken up when a blue jay or hedgehog photobombed the CCTV camera, but for the most part, what the videos showed was either nothing (all too often) or crickets fighting or mating.

If he wasn't in the field or in front of a monitor, Fisher spent time thinking about exactly what network metrics and which of the many network models available to use. For that, he did an even deeper dive into the literature, eventually finding some models in sociology he could modify and apply to the crickets.

Once all the tapes were finally transcribed, Fisher and his colleagues constructed both fighting and mating networks. The fighting networks they built showed that larger (heavier) males fought more opponents, and that males in general were more likely to fight with other males whose burrows were close to their own. While those two findings were fairly intuitive, another was not: Males who shared a common opponent in the power network around their burrows were more likely to fight one another than males who did not share a common enemy but who had burrows equally close to one another; that is, physical spacing per se did not explain what was going on. But it's even more subtle and complex than that. When Fisher and his colleagues looked at the cricket mating network, they found that males who had fought with one another were also more likely to mate with and successfully inseminate the same female than were males who had not fought: physical spacing per se did not explain that result, either. Who fights whom and who mates with whom are all entwined in a complex web of social networks in the field cricket reality show playing in the fields of that meadow in Spain.[7]

Temperatures in the meadow hosting that reality show rarely if ever dip below freezing at the time of year when the field crickets are active. There's also lots to eat, and while

After Hurricane Maria, rhesus macaques on the island of Cayo Santiago, Puerto Rico, increased the number of partners in their grooming networks, but the average strength of a grooming relationship did not change. Photo courtesy of Lauren J. N. Brent.

A group of sulphur-crested cockatoos feeding in Sydney, Australia. Color markings (as well as wing tags not shown here) are used to identify individuals. When food is relatively scarce, birds strengthen the bonds with others in their social network. Photo courtesy of the Wingtags Project.

Rock hyraxes on a cliff at the Ein Gedi Reserve in Israel. Rock hyraxes in egalitarian social networks live longer than in networks where a select few are central players. Photo courtesy of Eran Gissis.

Schools of manta rays at the water's surface. Manta rays form social networks at underwater feeding and cleaning stations in the Dampier Strait. Photo courtesy of Robert J. Y. Perryman (and his drone).

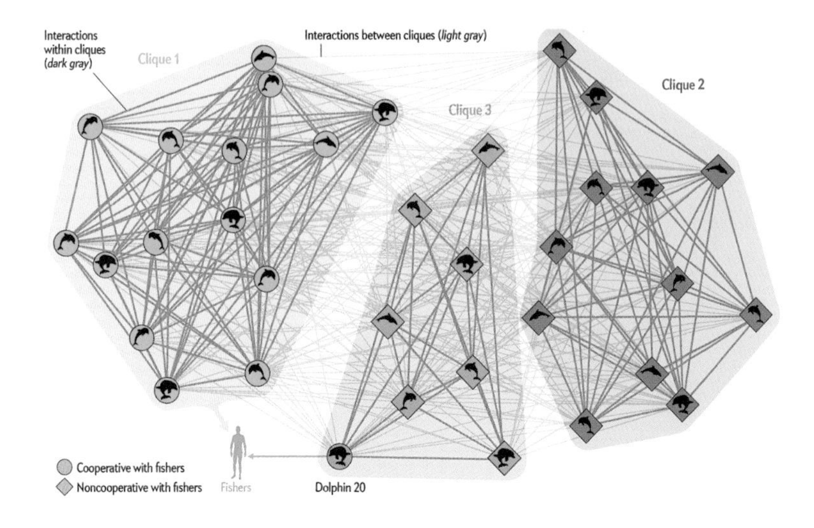

A dolphin social network in Brazil made up of three cliques. Circles = dolphins who signal fishermen, diamonds = dolphins who do not signal. All the dolphins in clique 1 (green) cooperate with local artisanal fishermen, who, along with the dolphins, are fishing for mullet. No dolphins in clique 2 (purple) cooperate with fishermen. In clique 3 (orange), only dolphin 20 cooperates with fishermen. From L. A. Dugatkin and M. Hasenjager, "The Networked Animal," *Scientific American* 312 (2015): 50–55. "Dolphins and Humans Team Up to Bag Fish." Reproduced with permission.

A group of house finches atop a cactus in the Sonoran Desert in the United States. Drabber males move between groups more often than colorful males, and in so doing game the network, using social lability to increase their reproductive success. Photo courtesy of Alex Badyaev, www.tenbestphotos.com.

Kangaroos in Sundown National Park, Australia. The more close friends a female kangaroo has in her foraging network, the lower her joey's chance of surviving to adulthood. Photo courtesy of Anne Goldizen.

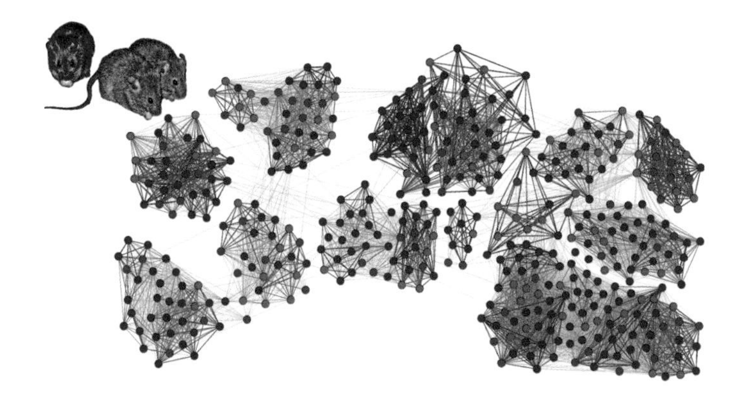

A house mouse network in Switzerland *before* a massive predation event. Red = mice found dead *after* the mass predation event, purple = individuals who went missing *after* the mass predation event, and blue = mice who survived the mass predation event. The thickness of a line indicates strength of association between mice. Mice with relatively few connections before the cat (or cats) struck formed many new, but weak bonds after the event. Mice who were well connected before the mass predation had fewer associates after the attack, but the bonds they formed were stronger than before the mass predation event. Figure courtesy of Julian Evans.

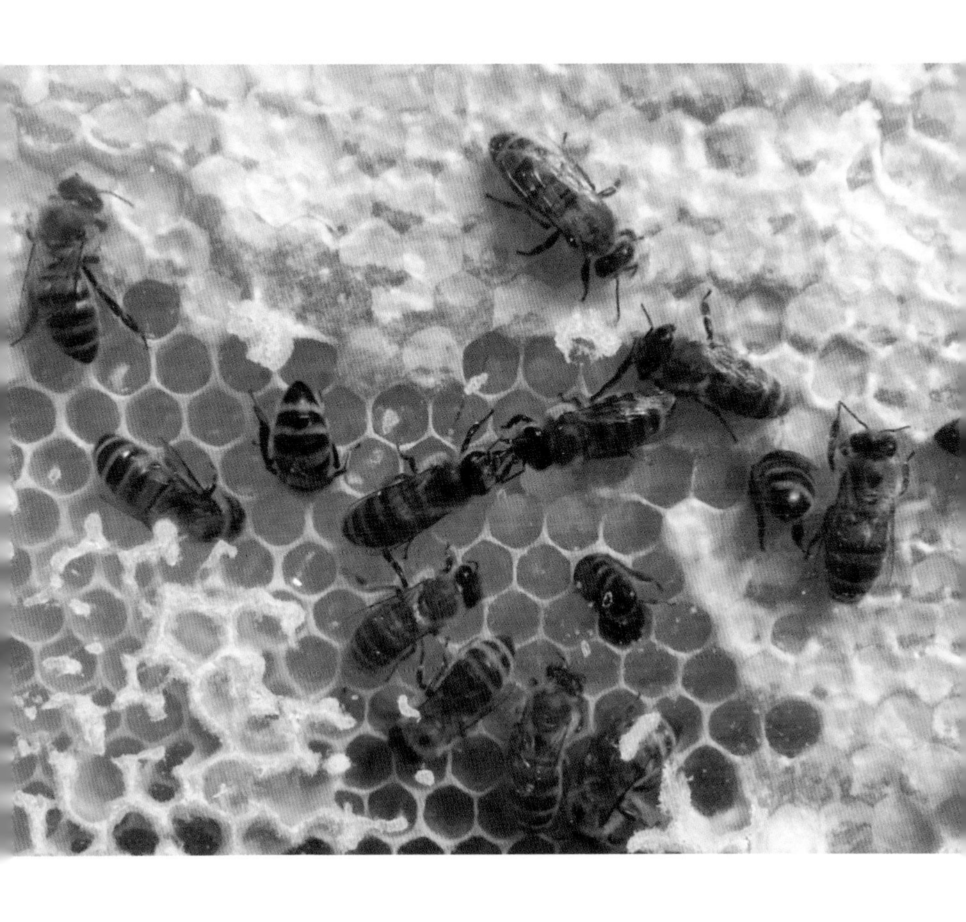

The two bees in the center of the photo are engaged in trophallaxis—the transfer of nectar from a forager to a bee at the hive. Trophallaxis networks are important when the position of the sun relative to the hive is unavailable to bees. Photo courtesy of Matthew Hasenjager.

Vampire bats in a roost at the Smithsonian Tropical Research Institute in Gamboa, Panama. In vampire bats' food-sharing networks, the amount of blood a starving bat received from member of its group was predicted by the amount of food it had given that individual when it was hungry. Photo courtesy of Dr. Simon Ripperger.

predators are always a problem for any population of animals anywhere, the meadow is a fairly cushy place to live most of the time. The mountains on the Isle of Rùm are not, and the feral goats living there have to deal with *that* reality.

Built on the Isle of Rùm off the west coast of Scotland in the last years of the reign of Queen Victoria, Kinloch Castle is still standing, albeit in need of repairs. Designed by the London architectural firm of Leeming & Leeming, the castle in its heyday was an exquisite hunting lodge for the family of textile magnate Sir George Bullough and their wealthy friends. Bullough's family owned the entire island, which lies 270 kilometers northwest of Glasgow, until the National Nature Reserves took control in the 1950s and turned it into a wildlife refuge. Today the red deer, Rùm ponies, Highland cattle, and Manx shearwaters on the island far outnumber the twenty or so permanent human residents.

For the last fifty years, ethologist Tim Clutton-Brock and his colleagues at Cambridge have been studying the red deer who live on the northern tip of Rùm. But it was the west coast of the island, with its rugged mountainous terrain, that drew evolutionary anthropologist Robin Dunbar to the island. Up in those mountains, a few hundred feral goats live on the edge of subsistence. One of the things Dunbar wanted to know was if networking helped them deal with just scraping by and, if so, how.

The odds are that if you have heard of Robin Dunbar, it's not for his work on feral goats (*Capra hircus*), but for a rather famous idea he developed in the early 1990s. "Thirty years ago," Dunbar wrote in a 2021 piece for *The Conversation*, "I was pondering a graph of primate group sizes plotted against the size of their brains: the larger the brain, the larger the group size. . . . Thus was born the 'social brain hypothesis.'"

Dunbar's work on the social brain hypothesis led him straight into the world of social networks. As he was pulling all these ideas together, Dunbar, who was at Cambridge at the time and in the same group as Clutton-Brock, was finding it harder and harder to get funding for studying primates in far-off places. His solution? "Like Tim [Clutton-Brock], I decided it was a much better strategy to work near home, and I came across the goats, by accident, because it was the only species nobody else [on Rùm] was working on." Soon Dunbar was telling anyone who would listen how intelligent goats were.[8]

The west side of Rùm is a harsh place for both goats and humans. Gales sweep off the Atlantic, and there are almost no trees to temper their bite during frigid winters. In the summer, the goats, as well as Dunbar and his team, were plagued by insects, particularly midges. Dunbar's group lived in a bare-bones nineteenth-century hunting lodge known as a bothy that has three rooms and a kitchen. Food was ferried to the island once a week. To pick up that food meant a 13-kilometer hike over mountains so rugged that a national rescue team trains there, over to the east side of the island where the Rùm National Nature Reserve office sits.

For the goats, who each day forage their way up 1,000-foot jutting sea cliffs, winter is especially brutal. It's not the cold per se, as they can buffer themselves a bit from the frigid temperatures by going back down the mountains and sleeping on the beach or, if they are lucky, in a cave. But food is another matter. "It is so bloody awful," says Dunbar. "For the goats, the real problem is time . . . because the day length is so short, that is what really hammers them. . . . If it is a cold winter, they just can't get enough food."

Dunbar was interested in how they coped and whether they used networking to deal with the challenges of life on Rùm. He recognized all the goats by coat color and natural markings and would just sit in the open, watching, as the goats didn't seem to care in the slightest. He'd peer at them using an old-fashioned admiral-style telescope, which he describes as "the kind [Admiral Horatio] Nelson used with the eye he could see through." He and his team took down data on power dynamics, what the goats ate, group size, distance between goats, and more.

It's a tricky question whether living in such a harsh environment should lead to peaceful group dynamics or ramp up the quest for power. On the one hand, goats are so time-stressed in the winter that they might need to devote all their time—every daylight minute—to foraging and therefore aggression might be rare. On the other hand, because every scrap of food is valuable, there might be power struggles aplenty for what little there is available. Fortunately, social network analysis can address this sort of question. When Dunbar and his colleague Christina Stanley plugged their data into a network model, they found the situation was far more interesting than what turned out to be a false dichotomy.

Dunbar and Stanley focused on two groups of females. When aggression took place inside these groups, it was most often of a rather mild sort, where one goat would move toward another, and before anything dangerous happened, the other goat would leave the area. On some occasions, though, such displacements included a nasty head butt from the approaching individual. Dunbar and Stanley first built a social network based on aggressive interactions, and then they built two more networks—one based on how close the animals were to each other when they were not in a contest, and the other based on "approach behavior," which involved one goat approaching another but, instead of displacement, the individual that was approached continued with whatever it happened to be doing at the time. These latter two networks captured prosocial behavior in goats, and analysis of each herd uncovered a core cluster of 12–13 females who made up the heart of all these networks, with others being outliers who interacted less often with one another and the core cluster.

On their first pass through all the networks, a positive correlation was found between the power network and both the approach behavior and proximity networks, which led to the counterintuitive implication that goats were aggressive with the very individuals they chose as friends. It turns out though that the goats were making subtle but important distinctions when it came to their frenemies. On a second pass through the network data, Dunbar and Stanley found that when the *strength* of the ties in the proximity network—that is, not just who a goat interacted with, but how often they interacted with the other goat—was compared to the *strength* of the ties in the power network, goats displayed less aggression toward those they spent the most time with. Dunbar was right about telling everyone how smart goats were from the get-go. On the mountains of Rùm, they don't just network; they distinguish between friends and acquaintances.[9]

It's one thing for goats to be living on the edge, on cold, inhospitable mountains, but that's not the sort of image we tend to conjure up when we think of primates. A humid rainforest in Africa or South America, yes, but a rugged landscape with snow is another matter altogether for primates. But for Barbary macaques (*Macaca sylvanus*) in the Middle Atlas mountains of Morocco, home, especially in the winter, is more like Rùm than Jane Goodall's Gombe. No one knows that better than ethologist Richard McFarland: "I set out to Morocco to study the general behavioral ecology of Barbary macaques," he says. "Then [one winter] snowfall wiped out their resources." Soon McFarland found himself trying to figure out how the power network in those macaques affected their very survival during that awful winter.

Four Barbary macaque groups, ranging in size from two to three dozen monkeys, live in the Atlas mountains' Ifrane National Park, 1,500-plus meters above sea level. Barbary macaques roam freely through the park, as do their predators, which include wild boars, genets, foxes, jackals, wolves, and eagles. Adult male Barbary macaques weigh in at about 16 kilograms and have pronounced, elongated canines that play a role in aggression, while females only tip the scales at about 9 kilograms and have much smaller canines. Starting around their fifth or sixth birthdays, males and females become sexually active, and breeding each year takes place in the fall–early winter with young being born in the spring. In the winter it can, and often does, dip below freezing, and even in the mildest of winters, snow covers the ground for about seventy days.[10]

The Barbary Macaque Project began in 2008 as the brainchild of ecologist Mohamed Qarro at the Ecole Nationale Forestière d'Ingénieurs in Salé, Morocco, and McFarland's PhD adviser at the University of Lincoln, primatologist Bonaventura Majolo. Majolo and Qarro found Barbary macaques particularly appealing, in part because they are the only macaques living outside of Asia and are listed as endangered by the International Union for Conservation of Nature (IUCN). What's more, very little was known about their behavior or, indeed, anything about them, and from a logistical standpoint, here was a population that primatologists-to-be graduate students in Europe could study without having to travel halfway around the world to do their fieldwork.

McFarland began his PhD in November 2008, at the very start of the project, and before he knew it, he was living full-time in Azrou, a town of 80,000, about 100 kilometers south of Fez and a thirty-minute drive to the study site in the mountains. Job one was to habituate the monkeys to his presence. When he started, if MacFarland got to within 100 meters of macaques, they just took off running. The first five months of his PhD research had him driving to the site every day and spending twelve hours "getting a bit closer, a bit closer to the point where they finally accepted [me]. . . . [I] never fed them or tried to stroke or touch them. . . . I sat still, walked around, and just showed no interest in them: I just became part of the furniture, a walking tree." Along the way, he learned how to recognize individual monkeys by their fur color, facial structures, scars, and whatever else distinguished one Barbary macaque from another.

After five months, the habituation process paid off to the point where MacFarland could get to within a few meters of the macaques. He would sit with his binoculars (to see the macaques who were farther away), and using a tablet, he would record data on aggressive interactions, group size, who was near whom, who groomed whom, what they ate, where they slept, and more. It was taxing work, as a troop of macaques can travel 10 kilometers a day in search of food. While most of the feeding was done on the ground, the long-distance traveling was done in the trees, where the macaques were moving much faster than McFarland could on terra firma. He did the best he could to keep up because he needed to follow the monkeys until they chose a tree to sleep in each night: if he didn't put a troop to bed, then when they awoke and began to move at the next sunrise, he might not find them at all that day.

It was hard enough keeping up with macaques in the summer, but in the winter snow it was brutal. "There were times I was deep in snow trying to keep up with these monkeys—

your feet are freezing, your fingertips are freezing," McFarland says. "You go home in the evening . . . shower, eat as much food as possible . . . and try to dry out your hiking boots next to a little gas fire."

After a year in the field habituating the animals to his presence, learning who was who, and developing an ethogram—a flow chart of 200 different possible macaque behaviors—McFarland felt more grounded. With a deeper sense of his study subjects, he assumed he could now go into high gear and get a true sense of what typical daily life was like for Barbary macaques. The problem was that right when this happened, in the winter of 2008/9, life was anything but typical for the Barbary macaques.

That winter was the coldest on record, with the ground covered in snow from November until late March. For the macaques, what that meant is that the seeds, lichens, fungi, leaves, and roots that they subsist on during the winter were buried under snow for five months. They were struggling to survive. "There was a time in January when every day another animal would be missing," says McFarland, "or I'd see a skeleton of a monkey that had been picked [cleaned] by a predator." By the end of the winter, thirty of the forty-seven Barbary macaques that McFarland and the Barbary Macaque Project team were tracking had starved to death.[11]

McFarland wanted to understand whether there was something special about the macaques who survived that brutal winter. Part of it was definitely skill—survivors were more skilled foragers—and part of it was probably luck: some found food when others had not. McFarland wondered whether group power dynamics might also have affected survival. As an animal behaviorist, he found that intriguing, but there was more to it than that. For someone who saw it as "a privilege just to have [these] animals indulge my presence," it was about understanding life and death in animals he cared for and knew like family.

McFarland and his adviser Majolo knew that social network analysis was a powerful tool for probing questions about group dynamics, but neither of them had any experience working with network models. But a colleague of theirs, Julia Lehmann, did. By the time McFarland and Majolo recruited her to help with the network modeling, Lehmann had done extensive work looking at social networks in chimpanzees and baboons. Together they built two different social networks based on behavioral interactions in the macaques in the six months just *prior* to that brutal weather. One was a prosociality network that was based on the number of times a pair of Barbary macaques groomed one another or were in body contact, and the second was a power network that used bites, slaps, full-on charges, grabs, and chases as its entries.

When they looked at *only* the prosociality network, they discovered that the number of ties that an individual had was positively correlated with survival, which would end up jibing well with results from a later study that found that in winter, macaques with lots of social partners during the day huddled together with more sleep partners at night. That sort of social thermoregulation reduces loss of body heat and likely increases survival.[12]

When McFarland and his colleagues looked at *only* the power network, they also found that the number of ties was positively correlated with survival. But there is an interesting twist to the story of power and survival. The power network analysis also found that macaques with a low clustering coefficient survived with a greater probability than others. Recall that cluster coefficients measure the extent to which an individual's neighbors are connected to one another. In a macaque power network, a low clustering coefficient means that a monkey is preferentially targeting aggression toward those who are not involved with many aggressive interactions with others. That makes good sense, but it would have never caught McFarland's eye without a network analysis.[13]

Once McFarland and his colleagues understood what each network alone told them about survival, what they did next was ask whether the power network, the prosociality network, or a combination of the networks *best* explained survival. The best-fitting model was the power network, and only the power network. In Barbary macaques, power bests prosociality; at least when its freezing, your food is covered by snow, and you are doing everything you can just to make it to the next day.[14]

Power networks matter. They matter for populations of insects in Spain, groups of birds in New Zealand, herds of goats on a small island in the United Kingdom, packs of monkeys in the Atlas mountains of Morocco, and much more. In extreme cases like that of the macaques, survival itself is at least partly determined by the dynamics of power networks. Still, a once-in-a-hundred-year frigid winter is, well, a once-in-a-hundred-year occurrence. But predators, and the dangers they pose, are something animals have to deal with every day: and as we are about see in everything from marmots to mice and zebras, network structure plays an important role when it comes to dealing with omnipresent threats to survival.

6. The Safety Network

I would give all my fame for a pot of ale and safety.
—Shakespeare, *Henry V*, III.2

In a now-classic paper, "Geometry of the Selfish Herd," evolutionary biologist William D. Hamilton built a series of mathematical models showing that natural selection favors grouping in the face of predation pressure, and that animals often jockey to get to the middle of a group, where it's safest. Hamilton was not the first to suggest that living in groups provides some level of protection, and since his paper, animal behaviorists have suggested many other benefits above and beyond getting to the middle of a selfish herd. These benefits include the dilution effect, in which larger groups dilute the probability that an individual is captured, and the "many eyes" effect, where larger group size increases the probability that some group member detects a predator before it attacks.[1]

Hundreds of papers have tested these ideas, but when it comes to dealing with danger, there's more to life in groups than diluting the odds, scrambling for position, and having lots of eyes on the lookout. For many species, there's networking.

To better understand the role that networks play in the

safety of network members, let's look in on mice dwelling in a special barn in Switzerland.

Twenty-five kilometers northwest of Zurich, atop a hill in a small patch of woods near the municipality of Illnau, sits a barn. From the outside, it looks like any other barn. But step inside the 800-square-foot building and you'd think someone had given a team of mouse city planners free rein to build four luxury rodent subdivisions. Instead, it was a team of ethologists, led by Barbara König, who did the designing and building and keeps up the maintenance. "I was interested in working with a wild mammal whose genome had been exposed to natural selection," König says. "In Europe, house mice always live in connection with humans. . . . They live in stables, barns, where we keep food." And that's been the case for a very long time: archaeological evidence suggests that house mice (*Mus musculus*) have been living in a commensal relationship with humans since 8000 BCE.[2]

Each of the mouse subdivisions in König's barn has ten cozy nest boxes, home to one or more females and their offspring, three water fountains to quench thirst, and ten mouse feeding stations serving a scrumptious mixture of oats and commercial rodent chow. Bricks and slabs of wood provide shelter and recreation. Any of the mice in the barn can move between subdivisions at will. Subdivision residents are also free to leave the barn through holes in the walls or under the roof, but if they do, they venture into a land of foxes, cats, badgers, and various species of very hungry birds. For the most part, they stay. But it's a hectic life. The mice are always on the move, sniffing for urine marks of others. "They don't," König says, "have the time that lab mice have."

König's work in the barn began in November 2002, when she and Andrea Weidt seeded the establishment with four male and eight female mice, all caught within 5 kilometers of their new home. Today there are around 400 mice in the barn, as well as a full-time research technician stationed there to make sure everything is under control. König and her team know the number of mice because they census every mouse in the barn, about every seven weeks. It's hard work and anything but glamorous. "All hands are on deck" for those censuses, says Julian Evans, who was a postdoctoral fellow on the research team: "We spend our days on our hands and knees emptying these nest boxes, and we do this amid all the grossness of hundreds of mice living in a barn."

To help automate the process of data collection, during censuses all mice weighing over 18 grams have a tiny PIT tag, weighing only a tenth of an ounce, inserted under their skin. The tag emits a unique signal that König and her team can read in a number of different ways. After a few years of tinkering, a powerful monitoring system was in place. Each nest box can only be entered through an acrylic tunnel that is bent at a 45-degree angle to slow down mice who zoom in. Two antennae are placed at the opening of each tunnel, and they pick up the signal of any mouse exiting or entering the nest. Antennae are also placed at drinking stations. All of this provides a nonstop stream of data on who's in the nest, when they come and go, and who they are interacting with outside the nest (at drinking stations). As if that isn't enough information, digital scales with infrared motion detectors are placed under some of the drinking stations so that the weight of mice can be updated on a regular basis.

This stream of data runs along cables buried under the sawdust floor of the barn into a laptop sitting on a table in the barn. At the end of each day, that information is automatically transmitted to a server at the University of Zurich, where it runs through a computer program that translates all that raw data into a form accessible to the house mouse team (entrances and exits from the nest box, stays in the boxes, etc.). Using all that data to build house mouse networks is where Evans, who joined König's team with experience modeling animal social networks, enters into the picture.[3]

When Evans joined the project, as long as the mice remained in the barn and did not venture out to the nearby surroundings, life was about as good as it could be for a rodent. Then came the weekend of January 19, 2019. The technician went in that Monday morning, Evans recalls, "and found carnage." A cat, maybe more than one, had somehow entered the barn, perhaps numerous times, over the weekend, and ravaged the population. Of the 478 PIT-tagged mice there the week prior, 85 lay dead in the barn, interspersed among another 32 untagged victims. Another hundred or so mice were missing, never to be found, despite an intensive search.

How a cat, perhaps many, got into the barn is not clear, but there are some clues. The patch of forest the barn sits in is surrounded by farmland, and there are a lot of farm cats nearby. "There were a few different cats we saw just hanging around at the back of the barn in this little hollow," Evans says, "just clearly waiting for a mouse to come out." In principle, the barn is supposed to be a sort of mouse fortress that is impenetrable, but as any cat owner will tell you, give a cat the smallest opening, and it can somehow squeeze through.

Mice have short generation times, and it wasn't all that long before the population recovered. But in the interim, König and Evans decided to make the best of a bad situation and explore how this large-scale predation event affected the social network of the survivors. A first pass through the data found that of the seventeen cliques that made up the full network in the barn before disaster struck, fourteen remained. Each and every one of those remaining cliques had lost members: mortality ranged from 12% in one clique to a devastating 88% in another.

König and Evans knew from other work that house mouse networks were complex, but they were about to find out just how complex. Think back for a moment to the macaques on Cayo Santiago and how they responded to the Hurricane Maria disaster. There, macaques used a simple rule to adjust their networks post-Maria: form more friendships, but don't strengthen already existing ones. House mice add a layer of complexity to the network rules for dealing with disaster. Mice who were connected to relatively few others in the network before the cat struck formed many new, but weak bonds after the event, perhaps seeking safety in numbers. Mice who were well connected before the mass predation event dealt with disaster in a very different way. They had fewer associates post-attack, but the bonds they formed were stronger than before the cat(s) came. Evans and his colleagues hypothesize that these mice may have become more "socially insular," but compensated for that by being involved in more prosocial interactions with their nest-box mates.

The house mouse team also looked at whether mass predation affected interactions between cliques. When they looked at clustering coefficients pre- and post-event, they found that cliques were less cliquey, with more movement of mice between cliques after the mass predation event. Cliques were a bit more porous, perhaps because after a traumatic event, being near any mice, even those outside your normal clique, provided a sense of safety. In any case, what seems clear is a that a *cat*astrophe at the population level can have profound effects on the structure of networks.[4]

Of late, König and her colleagues have become interested in the relationship between social network dynamics and the genes that are turning on and off in the brains of their mice.

But before we delve into that, let's move to the mountains of Colorado to see how marmots network to stay safe, and then we'll head to the Masai Mara National Reserve in Kenya, where impalas, blue wildebeests, zebras, giraffes, and eight other species network to fend off danger by eavesdropping on each other's alarm calls.

In the summer of 1919, John Johnson, a biology professor at Western Colorado College, headed off to do some fieldwork at a site near Gothic, Colorado, an abandoned silver mining town that sits in the Rockies at an elevation of a bit more than 2,700 meters. Johnson was struck by the diverse ecosystem, and soon he was bringing students there for field courses. Less than a decade later, in 1928, the Rocky Mountain Biological Laboratory was born; today it's a seventy-building complex that justifiably boasts that it hosts "one of the largest annual migrations of field biologists in the world." Until recently, researchers and students would stay in old mining cabins and other older housing units, but with support from the National Science Foundation, more modern housing is now available.[5]

In addition to all those humans, the Rocky Mountain Biological Laboratory is also host to many yellow-bellied marmots (*Marmota flaviventer*), a cat-size rodent in the squirrel family. These animals are a dream come true for an animal behaviorist. Marmots are out and about during the day, they reside in easy-to-find burrows, and have a yearly cycle of winter hibernation and wakefulness that maps nicely onto the academic calendar.

For the last two decades, the Rocky Mountain Biological Laboratory has been Dan Blumstein's second home and the yellow-bellied marmots his second family. Marmots live in smallish colonies comprised primarily of matrilineal groups—

females and their genetic relatives—plus one or two males, and Blumstein and an army of collaborators have been poking and probing the marmot social system. Depending on the year, Blumstein's team gathers information on sixty to more than 300 animals.

To sort out who's who, every marmot in the populations under study has a metal earring with an identification tag, as well as a temporary dye mark applied to its fur. When marmots wake up from hibernation, Blumstein and his team jump onto their skis and census the area, looking for animals emerging from their winter sleep. Starting in mid-April, they switch from skiing to either walking, biking, or driving. Using binoculars and spotting scopes, the crew perch themselves somewhere between 18 and 140 meters from the marmots and, using pen and paper, record what they see. They gather data on survival rates across years, as well as on friendly behaviors such as grooming, playing, greeting, sitting together, and social foraging. They also measure hormone levels from feces, blood, and fur samples that they collect during periodic trappings. Days can be long, but marmots can be very cute, and some of them are downright entertaining. Blumstein remembers one male in particular who lived to the age of eleven: "Some males are kind of brusque and other males are sweethearts. . . . This guy was always going up and greeting his wife. . . . Every time he would greet 1399 [the female he mated with], she would smack him in the nose."[6]

Fieldwork goes on seven days a week until about the second week of September, when the marmots call it a year and start slipping into hibernation. During that hibernation, Blumstein and his team are often using social network models to understand life in marmot society. Blumstein's sortie into social networks started in 2004 with his involvement in a series of workshops at the National Center for Ecological Analysis and Synthesis in Santa Barbara, California. These workshops were all about using evolutionary principles to help with national security matters, what Blumstein and others called "Darwinian homeland security": a "bunch of ecologists and evolutionary biologists and psychologists and security and defense people and warriors and other very interesting people" is how he describes those attending. As he prepared for the workshop, Blumstein found himself reading more and more articles on terrorism and terrorist *networks*. It didn't take long before he was applying network thinking to his marmots.[7]

Blumstein's thoughts soon turned to marmot network dynamics as they relate to safety against predators. Like other marmots he had studied, some yellow-bellied marmots give off alarm calls when a predator is detected. When others hear that call, they stop what they are doing and look around for danger, sometimes rushing to the safety of a burrow. Not all alarm calls are equal: some calls are noisier than others, and so likely alert more group mates. Blumstein and his colleague Holly Fuong wanted to know if position in a social network affected who gave off those louder alarm calls.

Every other week, Blumstein and Fuong set out traps near burrows baited with horse food, a marmot favorite. As they transferred the trapped marmot to a canvas bag, they had microphones placed about a foot away and recorded any alarm calls. Then they looked at whether an individual's position in a network that included many different friendly, as well as aggressive interactions, affected the type of alarm call produced. It did. More socially isolated marmots, who were involved in relatively few social interactions, gave off noisier calls. Blumstein and his colleagues think it may be that socially isolated marmots with few partners they can count on in the face of a putative threat are more easily aroused by danger.

Network dynamics affect not only how alarms are given, but how listeners respond to them. Though their results were not quite statistically significant, in a separate study, Fuong, Blumstein, and their colleague Elizabeth Palmer found hints that marmots with many friends and friends of friends began feeding more quickly than others after hearing an alarm call from an unfamiliar individual, perhaps because they feel more secure than marmots who were not as well connected. With an understanding of the power of the social network approach to unravel marmot social life, Blumstein and his team, led this time by Anita Pilar Montero, next turned to networks, predation, and mortality.[8]

Marmot mortality during the summer is almost always the result of predation, and so using data collected between 2002 and 2015, Blumstein's team mapped summer survival probabilities for both adults and one-year-olds (yearlings) onto networks built using all social interactions in that network. While no social network metrics were correlated with summer survival in adults, a telling sex difference appeared between male and female yearlings. For yearling males, none of the network metrics were correlated with summer survival, but yearling females with greater centrality measures were more likely to survive the summer than less-connected peers. Blumstein wanted to know why. In time, he came to think that dispersal behavior may hold the answer.

Leaving a group to try to make it on your own, at least until you find a new group you can join, is dangerous business for a marmot. Despite that, by the end of their yearling summer, almost all males disperse from the groups in which they were born, and that may explain why they invest less heavily in social interactions *before* they leave. But only about half of yearling females disperse. Blumstein knew that the young females who were more deeply embedded in their network were the ones staying put. When he pieced together the puzzle, he came to think that well-connected yearling homebody females were opting to avoid the dangers associated with dispersing to start life anew. And it paid off for them in one of the most important of all evolutionary currencies: survival.[9]

Sixteen thousand kilometers, give or take, separate yellow-bellied marmots in the Rocky Mountains and Jakob Bro-Jørgensen's field site in the Masai Mara National Reserve in Kenya. But it is more than distance that separates the studies, for Bro-Jørgensen's work adds a new dimension to our sortie into social networks. Up to this point, nodes in social networks have represented individuals, but they need not. In the work done at Masai Mara, the focus broadens, and nodes become species in a community.

In single-species networks, almost anything an individual does—feeds, grooms, selects a mate, fights, and so on—can impact, and is impacted by, others in the network. The situation is more complex when species are the nodes in a network. What individuals in species 1 eat may or may not impact foraging in species 2. Who individuals in species 1 select as mates virtually never affects mate selection in species 2. This means that the structure of multispecies networks needs to revolve around behaviors that are recognized and measured in the same currency across species. Because species in a community often face danger from a common set of predators, and some antipredator behaviors can serve as common currency, they may be especially important in structuring networked communities.[10]

Bro-Jørgensen did his PhD work at University College London studying topi antelope (*Damaliscus lunatus*) at Masai

Mara. Funded by the Zoological Society of London, he lived at the reserve from October 1998 through June 2000. Like all visitors to Masai Mara, Bro-Jørgensen would sometimes just stare in amazement, as a wide array of mammalian species paraded before his eyes.

Masai Mara National Reserve, which borders the Serengeti Reserve in Tanzania, covers about 1,500 square kilometers and is fed by the waters of the Sand, Talek, and Mara rivers. Depending on where you are in the reserve, you might see giraffes, hippos, waterbuck, warthogs, gazelles, eland, topi, zebras, baboons, black rhinos, and more. Between July and October, a million-plus blue wildebeests, as well as a hundred thousand or so Thomson's gazelles, migrate through the Serengeti-Mara system. Bro-Jørgensen saw all that diversity through the eyes of an animal behaviorist. "I was always looking around at these different constellations of mixed-species groups . . . but there were so many species there and it was hard to make sense [of it all]," Bro-Jørgensen says. "I thought it would be very interesting to look more into that."[11]

When he began to read more in depth, Bro-Jørgensen found that most research on multispecies interactions was being done by ecologists, and, as a consequence, it focused on resources in the environment rather than the social behavior of the animals. He thought that needed to change, and a good way to start would be studying the alarm calls that a wide variety of animals at Masai Mara emitted when a predator is sighted. But how? He knew the social network literature, and that almost all of it had individuals, not species, as nodes. That was in part because most multispecies communities of animals just didn't have enough species to allow for a powerful network analysis: but, Bro-Jørgensen thought, at Masai Mara that was not a problem. In September 2015, he got this chance to look at those multispecies networks when he and his colleagues, Kristine Meise and Daniel Franks, headed off to the reserve.

Bro-Jørgensen set his sights high. From his prior work at Masai Mara, he decided to gather antipredator behavior in each of the twelve most common herbivores that feed on the open plains: topi, Thomson's gazelles, common eland, Grant's gazelles, impalas, warthogs, hartebeests, blue wildebeests, plains zebras, African buffalo, giraffes, and even one bird species, the ostrich. At any given time, almost any combination of these species might be feeding together in a mixed herd. But individuals in each of the twelve species sometimes feed when they are not part of a mixed herd, which proved fortunate for Bro-Jørgensen and his colleagues, because data on antipredator behaviors, including alarm calls in single-species groups, could be used as a baseline to compare to the case when a species was part of a mixed-species herd.

To gather that baseline data for each of the twelve species, Bro-Jørgensen and his team first measured how vigilant individuals were when feeding only with members of their own species. They drove their Land Rover to within about 45 meters of a group and sat and watched, defining vigilance as the amount of time an animal was looking for predators rather than foraging. In time, they gathered data on 574 individuals across 109 groups in their twelve species.

Bro-Jørgensen and his colleagues next wanted information on alarm calls when individuals were in single-species foraging groups. But rather than wait for those rare occasions when they were lucky enough to be present when both a single-species group was foraging and when a predator—a lion, leopard, cheetah, spotted hyena, or black-backed jackal—was present, they took matters into their own hands, creating life-size posterboard models of each predator. The animals clearly perceived the model as a predator. And they weren't the only ones. "It was funny," Bro-Jørgensen says. "Sometimes we had tourists that came up [to the car] . . . and they wanted to see the big fights . . . and we would have them taking pictures of the lion. . . . We'd [eventually] say, sorry, but . . ."[12]

Before Bro-Jørgensen's team compared the vigilance and alarm-calling data from each species when in single-species groups versus when in mixed-species herds, they wanted one more bit of information: from the perspective of species 1, how salient was the alarm call of species 2? To figure that out, they first recorded the alarm calls made by individuals in each of their twelve species, and they played those recordings thousands of times using speakers they placed in the field. Next they looked at changes in vigilance in members of species 1 when exposed to the alarm calls of species 2, stepping through this for every pairwise combination of the twelve species. When they spotted individuals grazing in a single-species group, they drove up and put the speakers in place. Depending on the caller species and recipient species, animals ran away or raised their heads up to look—suggesting they processed the alarm calls from the signalers as true signs of danger—or they didn't respond at all.

Bro-Jørgensen's group used data from single-species groups combined with the information on who was grouping with whom in mixed species to construct multispecies alarm-call networks. What they found was that species whose members displayed high vigilance scores tended to be found in herds full of other species whose members were also quite vigilant. In these species, many networked eyes searching for, and many networked ears listening for, predators works at the community level. But there's a catch: species in which individuals gave relatively few alarm calls were especially likely to spend time as part of a mixed-species herd. That smacks of parasitism—having members of another species do the dangerous work—although Bro-Jørgensen is quick to note the data was not collected in a way to say that with any certainty. In any case, everyone likes to be part of a multispecies alarm-call network when others are prone to produce the calls: eavesdropping in networked communities pays off in all sorts of ways.

Understanding how a multispecies network functions suggests that what affects one species affects many. This means that the dynamics of social networks have conservation implications, particularly with respect to anthropogenic (human-caused) factors. Consider this: other work by Bro-Jørgensen and his colleagues has found that the relative importance (measured by centrality) that different species play in the Masai Mara herbivore community depends on ecological variables such as habitat structure. Habitat fragmentation by roads and fences is chopping up Masai Mara into pieces, which will affect how species interact when in multispecies communities. If a central species in the alarm-calling network was excluded, or even if its relative abundance was decreased, that could affect the safety of *all* the species in the network.[13]

Circling back to those mice in the barn outside Zurich that we met at the start of the chapter, what then is going on in their little but complex brains, at least as far as social networks and gene expression (i.e., what genes are turning on and turning off) are concerned? What Barbara König and her colleague Patricia Lopes did was measure the gene-expression levels of hundreds of genes expressed in three areas of the mouse brain. Then they looked for differences when comparing the gene-expression patterns of mice with many network ties versus few ties. They went in knowing that this approach would not allow them to figure out if differences in social interactions in the network led to differential gene expression or vice versa, but it was a start.

Males turned out to be fairly uninteresting on this front: across all brain regions, only three genes showed different expression levels as a function of the number of network ties a male had. In females, though, gene-expression levels differed rather dramatically between well-connected mice and their less-connected counterparts in the barn: 180 genes showed differences in expression levels across all brain regions.

A simple count-them-up approach is all well and good, but what König and Lopes really wanted was a sense of what those differentially expressed genes do. Focusing on the data they had from the hippocampus, they found that the number of ties a female had affected the expression of genes associated with chemical communication between neurons, via a structure known as the extracellular protein matrix (ECP). Experimental manipulation of the ECP by other research groups has shown it plays a role in learning and memory. We've already seen how learning plays a role in social network structure, and there is a lot of literature showing that mice learn about almost everything that might matter to mice: it stretches credulity to think that such learning doesn't play a role in house mouse networks. Gene-expression patterns may help us understand how.[14]

Whether it is a slew of herbivore species or a population of marmots or mice, how, and with whom, animals network affects how animals deal with putative danger. When putative danger becomes actual danger and death strikes, it can even restructure the social networks of survivors.

It makes perfect sense for networks to be important when it comes to antipredator behaviors, because animals are always on the lookout for predators. Another thing that animals are always doing is navigating around their environment. As we are about to see, some animals stay put more than others, some are leaders, and some are followers; all of that matters when it comes to the navigation networks they form.

7. The Travel Network

I saw there for the first time, carrier pigeons which take letters in their tail-feathers. — *The Journal of Pero Tafur* (ca. 1435–39)

Born circa 1410 in Andalusia, Pero Tafur had an incurable desire to travel. Between 1435 and 1496, he visited Genoa, Venice, Rome, Palestine, Egypt, Rhodes, Chios, Constantinople, and more. Of the many things he encountered along the way, Tafur made note of a rather special bird. Homing pigeons had been around long before he set sail in 1435, but Tafur had never seen one. That changed when he arrived in the Egyptian city of Damietta, where he saw them carrying letters in their tail feathers. "They carry them from the place where they are bred to other places, and when the letters are detached they are set free and return to their homes. By this means the inhabitants have speedy news of all who come and go by sea or land, and thus escape surprise, since they live without defences, and have neither walls nor fortresses."[1]

Animal behaviorists have long studied whether pigeons, including the homing pigeon (*Columba livia domestica*, likely what Tafur called a carrier pigeon), navigate by using the earth's magnetic field, the sun, key topographical features, sound, or some combination of these. They've also looked at the brain circuits and hormones that might play a role. But Andrea Flack's interest was different. Homing pigeons, like almost all breeds of pigeons, are group-living animals, and they're rather obsessive when it comes to navigating. They rely on particular routes and try very hard to take the same route every time. Homing pigeons also roost together and fly together in flocks. Where such sociality exists, networks are often in place. Flack was interested in whether networking played a role in how homing pigeons behaved in flocks returning to their roost. "If you have a [homing] pigeon and release it, and if there are other pigeons around, it will try to join them; it won't fly alone," Flack says. "How do they make decisions when they don't agree on where to fly? Do they make compromises? Does one become a leader, and, if so, what is this based upon?"[2]

Flack took on these questions as part of her PhD work with Dora Biro, a member of the Oxford Navigation Group (OxNav). The OxNav group had built two pigeon lofts that housed about 120 homing pigeons out in the Wytham Woods, which we first encountered in chapter 3. With the exception of days when experiments were underway, pigeons in the loft could come and go as they pleased, and often did just that, moving in and out of the loft in small flocks.

Flack and her team selected thirty pigeons from the loft and created three groups of ten birds each. Their protocol for studying homing behavior and social networks was elegantly simple. First, nothing happened unless the sun was shining because the birds use the sun to navigate. If the sun was cooperating, Flack would go to the loft, catch ten pigeons she had preselected, put them in a box, and drive to a release site. That site was exactly 15.1 kilometers from the loft, and all the pigeons had experience navigating back to the loft from sites that far away, but no experience from the release site itself.

Once Flack arrived at the release site, she took out the tiny GPS devices she had with her. Each GPS transmitter weighed less than half an ounce and had a small piece of Velcro on it. Flack then attached a device to a small Velcro patch glued to the feathers of each bird. The ten birds were then released as a group. The procedure was repeated on multiple days, until each of three groups had been released from the release site and allowed to (hopefully) find their way home on eight separate occasions. Between releases the GPS transmitters were removed. "It is *very* dirty in the pigeon loft," says Flack. The birds would likely peck off each other's GPS devices, and then "they'll just poop on it."

The homing pigeons lived up to expectations. Fifteen kilometers proved little challenge in terms of them finding their way back to the roost. It usually took the birds ten minutes or so to fly back to the roost, and they often got there before Flack could drive back. On occasion a pigeon might get lost, but that was rare. And even less often, but still problematic when you only have so many pigeons to work with, a predator might kill a bird as it flew back to the loft. Flack never saw that with her own eyes, but the GPS data told the story for her: when they were flying together in a compact flock, and all of a sudden birds start flying in different directions, she knew *something* had happened. If one or two birds never returned to the loft, she assumed that something was a predator catching one of the pigeons.

To probe the role of social networks and the role of possible leaders and followers in those networks, Flack took the GPS coordinates for each and every pair of pigeons in a flock, and she and her colleagues looked at how often bird A was leading and bird B was following A—by responding to a directional change—as they flew back to the loft. When viewing the data for an entire flock, "networks really show up nicely," Flack notes. A clear leadership hierarchy emerged in each of the three flocks with birds at the top of that hierarchy—the leaders—having a disproportionate effect on directing the group as it flew home.[3]

Flack's interest in leadership networks and travel only grew after her dissertation was complete. Animal behaviorists, including Flack, are a plucky lot, always up for a challenge. Even so, many would say that moving from studying leadership networks in pigeons navigating 15 kilometers to looking at leadership networks in flocks of European white storks (*Ciconia ciconia*) as they fly along at a clip of 100 kilometers an hour, soaring 1,200 meters above the ground and migrating thousands of kilometers from across Europe to various locations in Africa, was a big leap. Fortunately, Flack, Martin Wikelski, and their colleagues at the Max Planck Institute for Ornithology—Flack's next stop on her academic journey as an ethologist—did not think so. They wanted to know if leader-follower networks were at play in larger-scale movements, and the migration of white storks was a perfect place to look.[4]

Before we join those stork leaders and followers on their monumental migration, let's first plant ourselves back on the ground and peek in on Tibetan macaques in the Mount Huangshan Biosphere Reserve of eastern China and northern muriquis in the Feliciano Miguel Abdala Private Natural Heritage Reserve in the Brazilian network as they travel about in their networked societies.

The Huangshan mountains were named by Tang Dynasty Emperor Xuanzong almost thirteen centuries ago, in 747, the same year that legend has it the elixir of immortality was found somewhere among its dozens of imposing peaks. Soon hermits, poets, and artists came to Huangshan. Sixty-four temples, many for the practice of meditation, were built on the mountain during the Yuan Dynasty (1271–1368); twenty of those temples still exist today. Many centuries later, the Shanshui (mountain and water) school of landscape painting was born there. Today the Mount Huangshan Biosphere Reserve, about 600 kilometers southeast of Beijing, is a UNESCO World Heritage Site and is home to more than 2,000 species of plants and 417 different species of vertebrates, including clouded leopards (*Neofelis nebulosa*), Oriental storks (*Ciconia boyciana*), and Tibetan macaques (*Macaca thibetana*).[5]

The Tibetan macaques live in an area of the mountain range called the Valley of the Wild Monkeys. There they move about in deciduous and evergreen forests, where they feed primarily on bamboo, grass, fruits, and tubers, although they can, and do, eat bark when other food is scarce. During the summer they spend the nights in trees, where danger from venomous snakes is minimal, while in the winter they prefer huddling together on the ledges of rocky cliffs.

Tibetan macaques can live to age thirty, and they spend much of that time involved in some sort of social interaction: thirty-three different social behaviors—including play, grooming, embracing, and teeth chattering—have been documented in the populations of the Valley of the Wild Monkeys. In 1986 Jin-Hua Li and Qishan Wang began studying the Yulingkeng (YA1) population of monkeys in this valley. Over the subsequent decades, Dong-Po Xia, Amanda Rowe, Gregory Fratellone, and more than 150 researchers and students from China, Japan, the United States, Australia, England, and Germany have studied anything and everything Tibetan macaque.[6]

By the time Dong-Po Xia joined the macaque research squad in the early 2000s, there was a field station with cabins where he could stay, and the Tibetan macaques were long since used to having humans poking around. Part of that was due to tourism in the reserve, but more to the point, the macaques were (and are) provisioned four times a day by rangers at sites far away from where Xia and others work. The YA troop lived just a few minutes' walk from Xia's cabin, and each morning at about 6 a.m., armed with paper, pencil, and digital tape recorder, he would head out to study twenty-four macaques in that troop. Xia could recognize individuals from natural characteristics and is quick to note that "we get the behavior data in the forest, not in the provisioned area." He gathered that data until about 6 p.m., when he'd follow the troop to their sleeping site for the night, so that the next morning he'd know where to find them.

Xia's dissertation work focused on social grooming networks in the YA troop. He was especially keen on using network analysis to study grooming relationships among females. In Tibetan macaques, when males reach sexual maturity at age six or seven, they disperse from their natal group and join, or try to join, another group. Females live their entire lives in the group in which they were born. This has two effects that might be important in social grooming networks. For one, adult females in a group have more experience with one another than do adult males, and for another, adult females in a group are often genetically related to one another, but that is rarely the case for adult males. Because female sociality lies at the heart of life in groups of Tibetan macaques, Xia and his colleagues predicted that adult females would have higher centrality scores—more friends and friends of friends—than adult males in grooming networks and that females would form cliques within grooming networks.

Between May 2009 and August 2010, Xia spent thirteen months living in the reserve, each day going out to the forest and gathering data on grooming relationships. He'd select an animal, and then for the next twenty minutes, while standing about 7 meters away from the monkey of interest, he'd take notes on whether it groomed another YA troop member and whether others groomed it. When the data from the 760 hours of such observations were plugged into a social network model, adult females did indeed have centrality measures that were greater than those of adult males. Xia's analysis also found five cliques and, with the sole exception of a single adult male in one clique, all clique members were adult females.

Xia's interest in Tibetan macaque social networks led him to join another project, this one headed by Amanda Rowe, who was conducting a study on travel networks in the YA troop. That work focused on leaders and followers, the latter of whom Rowe refers to as "fans." Perched on platforms that allowed them to look down on the troop and track movement, Rowe, Xia, and their colleagues set to work. They defined an act of leadership as a monkey moving at least 10 meters from its otherwise stationary group, with two or more other macaques (fans/followers) joining it within five minutes. Network analysis found that dominant females were leaders, and so had more fans than other macaques. In a twist, Rowe and her team found that because dominant females were often genetic relatives of other dominant females, in addition to having more fans, dominant females themselves were also fans of more network members (their dominant relatives) than lower-ranking macaques.

Gregory Fratellone wanted to know even more about macaque travel networks and how they are tied to grooming networks in the Valley of the Wild Monkeys. As part of his master's thesis, he, along with his colleagues, thought to combine information on centrality in grooming networks with the speed at which travel groups coalesced for a bout of exploration. What they found was that there were four YA travel cliques in the Valley of the Wild Monkeys. Cliques varied in size, as well as in sex ratio. Not surprisingly, it took smaller cliques less to time to come together to prepare for exploration than it did larger cliques. And, as with almost all things Tibetan macaque, female sociality reigned supreme. When Fratellone's team looked at the time it took a travel clique to jell as a function of the sex ratio in a clique, they found that cliques with a greater proportion of females were in place and ready to explore more quickly.

Part of the reason females are especially adept at coalescing into travel cliques quickly is that they were well connected via the grooming network, so that they knew each other well and had built up bonds of trust. Still, there's likely more to it than just that. Other work in the YA troops has found that dominant males sometimes redirect their anger at each other toward infants. Mothers need to be ready to leave quickly and explore new options when that happens, and when they do, it may be with their female kin, who also have a genetic stake in the welfare of their relative's infant.[7]

For Tibetan macaques, travel networks are about clique formation *within* a population. In primates, though, sometimes populations fuse and sometimes they divide, giving birth to new groups. For the northern muriqui (*Brachyteles hypoxanthus*), a monkey who calls the forests of Brazil home, that group birthing process is intimately tied to the structure of travel networks.

Brazilian animal behaviorist and conservationist Marcos Tokuda has wanted to study monkeys in the wild for as long as he can remember. To make that dream a reality, after finishing his undergraduate degree in the early 2000s, he contacted anthropologist Karen Strier at the University of Wisconsin. Strier had been studying northern muriquis in the forests of Brazil since 1982 and had recently published a monograph, *Faces in the Forest: The Endangered Muriqui Monkeys of Bra-*

zil. Strier told Tokuda that Jean Philippe Boubli, a postdoctoral fellow in her lab, had some funding from the National Geographic Society and was looking for a student to help him study northern muriquis. Tokuda jumped at the opportunity. In short order, he had added Patricia Izar, a primatologist who had worked with social network theory, to his team of mentors and entered the graduate program at the Universidade de São Paulo, where Izar was a professor in the psychology department.[8]

Tipping the scales at about 9 kilograms and standing about 60 centimeters tall, northern muriquis have long prehensile tails that are especially useful for navigating through the forest canopy. Their faces are hairless but carry a lovely combination of pink and black, and they have a rudimentary thumb that comes in handy when they gather fruit and seeds. The troops that Tokuda studied for his thesis roamed the forest of the Feliciano Miguel Abdala Private Natural Heritage Reserve in the Brazilian state of Minas Gerais, on the southeastern end of the country about 200 kilometers west of the Atlantic Ocean and 1,200 kilometers northeast of Rio de Janeiro. The IUCN Red List categorizes northern muriquis as "critically endangered," with an adult population worldwide of a thousand and falling. About 20% of those thousand live in the Abdala Reserve.

Some of the animals that the northern muriquis shares the forest with in the Abdala Reserve are buffy-headed marmosets (*Callithrix flaviceps*), ocelots (*Leopardus pardalis*), and Ruschi's rats (*Abrawayaomys ruschii*), and vinaceous-breasted amazon parrots (*Amazona vinacea*). It is a gorgeous place to work, but not without its challenges: brutally hot and humid during the day, with ticks everywhere. And it's isolated. "Being so far from your family and friends," says Tokuda, "you need to be mentally tough."[9]

That mental toughness involved living at the reserve for eighteen months, with only brief visits back home to São Paulo every couple of months. Each morning Tokuda would get up at 4 a.m. and head out to the forest, where he would spend the entire day tracking troops of northern muriquis. The monkeys are not marked or tagged, and so for the first six weeks in the field, Tokuda devoted his days to sketching and jotting down details of the facial color patterns of individual monkeys so he could recognize each animal.

Shortly before Tokuda joined the project, a group fission began, in which the Jaó troop of monkeys was splitting into two, forming a new troop that was dubbed the Nadir troop. As Tokuda was learning about the individual identities of the monkeys, and then as he started to gather behavioral data, he couldn't help but be fascinated with the behavioral dynamics of the fission underway, as he watched some individuals moving from the Jaó troop to form their new Nadir troop on a territory whose center lay a bit over a kilometer south from the center of the Jaó territory.[10]

Each day Tokuda would do "roll calls," in which the identity of every monkey in the Jaó and nascent Nadir troop was noted. He also collected information on who was traveling with whom and, using a handheld device that gave him GPS coordinates, where the monkeys traveled. In time he ran 820 roll calls and collected 11,300 GPS measurements, spanning two wet seasons and two dry seasons at the reserve.

The fissioning process was gradual and took about a year, but once it was complete, the Jaó troop had thirty (or so) adults or subadults, and the new Nadir troop had settled down to a size of about forty. But it was the dynamics of the fission process, not the size of the troops, that most interested Tokuda and his colleagues, and social network analysis was the perfect tool to probe those dynamics. Before they even began calculating network measurements, Tokuda's team created visuals snapshots of the network, and then networks, over the course of the fissioning process. Those visuals left no doubt, as they could see the initial, pre-fissioned Jaó network slowly but surely pulling apart into the Jaó and Nadir networks.

What happened became even clearer with a network analysis of movement and association patterns. The fissioning had been initiated by females, who in time were joined by males. But it was not a random draw which of the females got the fissioning process underway. Females who were weakly connected to one another in the original Jaó network—those who had low strength and low centrality scores—were the ones who left to establish the Nadir group. It's rare to see the birth of a new network in the wild in real time and even rarer to be able to piece together how dynamics in an ancestral network fundamentally affect the formation of its "offspring" network. But for northern muriquis, thanks to Tokuda and his colleagues, we now know at least one way this fissioning can play out: when a subset of females sense their connection to others in a network has dipped below some minimum threshold, they opt to leave and start their own troop, with males eventually following.[11]

When pigeons, Tibetan macaques, and northern muriquis network to travel, they're moving a couple of dozen kilometers. The white storks that Andrea Flack, whom we met earlier, studies take travel networks to another level.

About 10 kilometers from the Max Planck Institute for Ornithology in Seewiesen, Germany, sits what Flack calls a "stork village" that is home to about two dozen nests. During the summer of 2014, she and her colleagues attached solar-powered GPS loggers to all sixty-one young juvenile white storks born in that village to track the behavior of their migration south, in part to see whether a leader-follower social network was present. Attaching GPS devices was a bit trickier for the storks than the pigeons Flack had worked with, but once the devices were on, they transmitted data on where the storks were, their relative position in a flock, and the speed at which birds were moving in all three dimensions (forward, to the side, and up and down). Flack and her team could keep track of that data on an app called Animal Tracker.[12]

A group of twenty-seven juvenile storks that Flack was tracking left to migrate south together in the autumn of 2014. Some of the storks with GPS devices veered off and joined different flocks, leaving a GPS-tagged group of seventeen birds. Even so, when she and her team tracked the first five days of that sixteen-bird migration, which brought the birds down to southern Spain, the GPS devices had transmitted a whopping 1,414,226 location coordinates.

Those seventeen juvenile storks were not migrating alone: they were always part of a larger flock (not fitted with GPS devices). What's more, Flack was also along for the ride. Storks don't migrate at night, and so each evening at about 7 p.m. they descend, most often avoiding areas trafficked by humans, and roost for the night. "The GPS data told me where they were, and I actually went and migrated with them," she says. "Usually I would get the data [in the evening] and then drive 300 kilometers to find them in the middle of the night, so I could watch them in the morning when they took off again." Flack slept in her car most nights, and conditions, at least for a human, were usually not ideal, including stretches on a Spanish pig farm. But it was worth it, as the early morning observations gave her estimates of the total size of the flocks her GPS-tagged birds were flying in.

As with many migrants who travel long distances, white storks are always looking to locate thermal uplifts, which allow for low-cost soaring instead of high-cost wing flapping. These uplifts are caused by thermal convection and what's called orographic uplift, which is the result of the deflection of horizontal wind as it passes over hills and ridges. Use of those uplifts turned out to be key to the navigation leadership network that emerged in the storks.[13]

Using the same measure she had used for the homing pigeons back in Oxford, Flack examined whether there were leaders and followers in the juvenile storks. During the course of the first five days of the migration, each of the seventeen GPS-tagged birds was at various times at the front, in the middle, or at the rear of a group, but overall a leadership network emerged with seven birds being most often at the front of the group, leading the others. Leaders spent more time soaring and less time wing flapping than followers, as they were especially good at locating uplifts. But uplifts are very dynamic, drifting with the wind, and the best location to circle within an uplift is always changing. To compensate, leaders showed movement patterns that suggest they were constantly reassessing where that ever-changing best location was at any given moment.

It's all well and good that the leadership networks emerge as the storks migrate south. But do those networks matter? Do leaders reap benefits from taking on their roles? Flack found they do. How far a bird migrates is a function of how much time it spends soaring versus wing flapping, and because leaders soar more and flap less than followers, on average they migrated farther south on those first five days of their journey. Many of the followers ended their migrations in Spain, but most leaders kept going, eventually making their way down to Africa, where they remained until the return migration north in the spring.

Travel networks, with their leaders and followers, dictate much about how individuals and groups move about within and between environments. And when animals move, as well as when they stay put, they communicate with one another about so many things. The information they communicate moves along networks. To see how this works, we'll start by spending some time with chimpanzees, who know a thing or two about communication.

8. The Communication Network

He learnt to communicate with birds and discovered that their conversation was fantastically boring. It was all to do with wind speed, wing spans, power-to-weight ratios and a fair bit about berries. — Douglas Adams, *Life, the Universe, and Everything* (1982)

A three-hour drive northwest from Uganda's capital of Kampala, Budongo Forest, at an altitude of 1,100 meters, sits on an escarpment that directs the Waisoke, the Sonso, the Kamirambwa, and the Siba rivers that run through down to Lake Albert. Average annual rainfall in the forest is about 160 centimeters, and the forest is home to more than 300 species of birds, 400 species of butterflies and moths, and 24 species of mammals, including 9 species of primates. In 1962 British biological anthropologist Vernon Reynolds began the Budongo Forest Project, a long-term study of 800 or so chimpanzees (*Pan troglodytes schweinfurthii*) living in the forest.[1]

For two decades, primatologists Anna Roberts and Sam Roberts have been studying communication networks in one community of chimpanzees at Budongo. For her dissertation work, Anna Roberts worked with the six dozen or so chimpanzees who make up the Sonso community, one of six chimpanzee communities in the forest. Roberts is interested in cognition and communication in nonhumans and thought it best to work with our closest evolutionary relative. And because it was already known that non-vocal gestures played an important role in chimpanzee communication, she focused on those. Sam Roberts couldn't agree more about the importance of non-vocal gestures in chimpanzees. "Living in groups has benefits . . . but also comes with many challenges," he says. "Gestural communication [is] a modality that can be used to manage relationships at a very fine-grained level."

On most of her visits to the forest, Anna Roberts stayed at the Budongo Conservation Field Station, with its housing for visiting researchers, a cafeteria, an herbarium, and even a small library. The Sonso chimpanzee community is a short walk to the north, and each morning she would head out, accompanied by two Ugandan field assistants. The chimpanzee community as a whole usually has about seventy members, with individuals moving about in groups of five or so and the composition of group membership sometimes changing day to day.[2]

Each morning, Roberts would head out to the nearby bush. "Chimpanzees are good study subjects because they don't travel that much," she says. "I also study bonobos and they travel like every twenty minutes... but once chimpanzees find a good feeding [site], they can be there all day." That allowed for lots of filming, and armed with a video camera, she would select a chimpanzee and follow him or her for eighteen minutes. Geresomu Muhumuza, one of her field assistants who has been involved with work on the Sonso community since 1990, is exceptionally talented at recognizing individual chimpanzees by their unique facial and body features, and while Roberts had the camera focused on her subject, Muhumuza took data on which chimpanzees were within 10 meters of the subject and what those individuals were doing. Then Roberts would pick a different chimp and start all over.

Over the course of three years, as Muhumuza gathered information on who was in the vicinity, Roberts filmed nearly 5,000 gestures made by Sonso chimpanzees. Gestural communication was defined rather strictly as the use of the limbs, body, or head in an intentional manner and given in the presence of an audience. A gesture had to be given in the direction of audience when the audience was oriented toward the signaler, and it had to lead to a predictable change in the audience's behavior and for a gesture to be considered intentional, it had to be repeated when it failed to elicit the typical behavioral response but stopped once the audience responded.[3]

Roberts and her team identified an astounding 120 different gestures in the Sonso chimpanzees, including the visual, tactile, and auditory gestures in table 1. Tactile gestures were almost always tied to friendly behaviors, such as grooming and playing, both of which reduce stress. Auditory and visual gestures were linked to friendly behaviors like greetings, grooming, playing, copulation, travel, and food sharing, but also to nasty behaviors such as threats and even more aggressive actions.

Little enough was known about the details of many of these gestures that an inventory, along with notes on possible function, was valuable to the animal behavior community. But Anna Roberts wanted to do more than just provide a laundry list of chimpanzee gestures at Budongo; she wanted to understand how those gestures structured chimpanzee societies, and one way to do that was by using social network analysis. That's where Sam Roberts's experience with network analyses came in handy. He had finished his PhD a bit before Anna finished hers and headed to the University of Sterling, where

Gesture name	Type of gesture	Action
Arm beckon	Visual	Fast sweeping movement toward self in horizontal plane
Roll over	Visual	Lying on surface, whole body rotates on x-axis
Bow	Visual	Upper back bends forward from the waist
Kiss	Tactile	Closed mouth brought into short contact with another's body
Embrace	Tactile	Arms placed around the torso of another
Hold hands	Tactile	Clasp each other by the hand
Tap object	Auditory	Strike object with inner side of open hand, slight and rapid
Lip smack	Auditory	Lips make loud sound by clapping upper and lower lips together
Drum	Auditory	Feet brought into short, audible contact with the buttress while holding onto a tree trunk

Note: Column 3 is quoted verbatim from A. I. Roberts, S. G. B. Roberts, and S. J. Vick, "The Repertoire and Intentionality of Gestural Communication in Wild Chimpanzee," *Animal Cognition* 17 (2014): 317–36.

Anna was. While he was there, he landed a postdoctoral fellowship with Robin Dunbar (whom we met in chapter 5). As part of that postdoc, Sam was looking at social networks in humans, and it struck him that this was also the perfect tool for looking at the role of gestures in chimpanzee groups.

But before we dig into communication networks in our closest evolutionary relatives, let's step back and explore networks in two species that aren't nearly as closely related to humans: the cosmopolitan honeybee and the silvereye (a songbird found across Australia).

Coordinating foraging behavior in a hive of thousands of honeybees (*Apis mellifera*) is no easy feat. Hundreds of workers, all female, buzz about looking for pollen and nectar wherever they can find it. What they gather, they bring back to the hive. This system poses logistical nightmares for the bees, including how to communicate information about food sources so that more workers can join in and bring back even more booty to the hive. That problem is solved, in part, by the waggle dance.

The dance behavior of honeybees returning from a bout of foraging was brought to fame by Karl von Frisch, whose decoding of dances was one reason he shared a Nobel Prize with Nikolaas Tinbergen and Konrad Lorenz in 1973. When pollen and nectar are far from a hive, sometimes up to 10 kilometers away—about half a million bee body lengths—foragers start waggle dancing upon their return home. Thomas Seeley, an animal behaviorist who specializes in the social behavior of bees, calls the waggle dance "a unique form of behavior in which a bee, deep inside her colony's nest, performs a miniaturized reenactment of her recent journey to a patch of flowers. Bees following these dances learn the distance, direction, and odors of these flowers and can translate this information into a flight to specified flowers. . . . The waggle dance is a truly symbolic message, one which is separated in space and time from both the actions on which it is based and the behaviors it will guide."[4]

To see how the waggle dance plays out in a hive, imagine that a worker bee has just returned from a distant patch of flowers full of pollen and dripping with nectar. Suppose that the patch of flowers is about 200 meters from the hive, 40 degrees west of an imaginary straight line running between the hive and the sun. When a forager returns to the hive, she starts dancing up and down a vertical honeycomb within the hive, waggling her abdomen as she does, with her sister workers in physical contact with both her and each other in the process.

While dancing, a forager conveys topographical information about the food source from which she has just returned. Compared to a straight up-and-down run along a honeybee comb, the angle at which a returning female dances (40 degrees, in our case) provides information about the position of the food source of interest in relation to the hive and to the sun. The longer a bee dances during one particular section of the waggle dance, the farther away the bounty: every 75 milliseconds of dancing translate into the resource being about an additional 100 meters from the hive.[5]

In addition to picking up information from the waggle dance of successful foragers, would-be foragers at the hive can also learn about food sources by the odor they pick up when their successful sisters return to the nest, either when nectar is transferred from a forager to another bee at the hive (trophallaxis) or when a forager rubs its antennae with a hive mate (antennation). Matt Hasenjager and Elli Leadbeater have been looking at communication networks as they pertain to waggle dances, trophallaxis, and antennation.

When I last visited Matt Hasenjager, he was a postdoctoral fellow at Royal Holloway, University of London, working in Leadbeater's lab on communication networks in honeybees, but it wasn't long before he and I took to reminiscing about the days when he was doing his PhD in my lab at the University of Louisville. Right from the start, Matt had been keen on delving into the dynamics of social behavior. At the time my lab was stocked with plenty of guppies (*Poecilia reticulata*), all descendants of fish caught during various trips to Trinidad years earlier. He dove in and designed an experiment to examine how social network structure affected learning and finding food in these fish. To do that, Matt created schools of guppies that differed in how many bold versus shy fish they contained, and then looked at how familiarity with other group members and the ambient threat of predation affected the flow of information about feeding sites. As he ran those experiments, and then watched the seemingly endless videotapes they produced, Matt somehow found the time to master the conceptual and analytical underpinnings of social network theory, with some significant help from Will Hoppitt, a guru in the mathematics of social networks in nonhumans.[6]

As a postdoc at Royal Holloway, Matt and Elli Leadbeater teamed up with Hoppitt to use social network analysis to examine what sort of information workers at the hive were picking up from foragers who had just returned with a haul of food. Their working hypothesis was that if workers learned about a feeding site through information conveyed in a waggle dance, then that information should diffuse along social networks associated with those dances. To test that, they used a model that Hoppitt had designed. That model looked at whether waggle dance networks predicted the order in which honeybees arrived at foraging sites. To examine the role of odor in learning about such sites, Hasenjager and his team also looked at trophallaxis networks and antennation networks in the bees.

The experimental honeybee hives at Royal Holloway sit in a room in a building at the center of campus. All interactions in a hive that can include upward of 3,000 bees are videotaped using strategically placed video cameras. Each hive has a tunnel that leads to the outside, and bees are free to come and go as they please. Or at least they were after an initial human glitch, that caused "a whole traffic jam at the entrance of the hive," as Matt tells it, was corrected.

Matt and his colleagues set up two experimental feeders on campus that provided the bees with sucrose, a honeybee favorite food. They used a tried-and-true method, employed since the early days of von Frisch's discovery of the waggle dance, to get to the point where the experiment could start. Bees were trained to the presence of the two feeders by placing them a short 10 meters from the hive, at an angle of 110 degrees relative to each other and the hive. As bees arrived at a feeder, a member of the honeybee team—usually Matt—would paint a tiny number on their back. That wasn't as difficult as it sounds as foragers seemed laser focused on gathering sucrose. And there was something about those enamel numbers that changed the human-bee dynamic. "As soon as you marked one of the bees," Matt says, "you felt like that bee suddenly became an individual. . . . When you have 3,000 unmarked bees that more or less all look the same, it's [different]. . . . Now, you're like 'Oh yeah, 223—you're one of my reliable bees.'" In any case, after bees were marked, it was easy enough for Matt or one of his colleagues to confirm whether or not they were members of the experimental hive and to determine which hive members specialized on which feeder.

Over the course of a few days, Matt moved the two feeders farther and farther from the hive (but always at the same angle relative to each other and the hive), until they were a 200-meter flight from where the bees lived. At that point, the experimental manipulations could get going, weather permitting. "We were really trying to target those nice, sunny, calm days," Matt says with a smile, "but it was England, so it didn't happen as often as one might want." Each of the two feeders was filled with the same amount of sucrose. Then, one day before the trial, one of the two feeders was emptied, while the other remained fully stocked. Now foragers, who day after day had specialized on what, all of a sudden, was an empty feeder, were "unemployed," a lovely descriptor that Matt and his colleagues use. What that meant was that the honeybee team could look at the relationship between foragers who specialized on the full feeder, and unemployed workers, who were now potential recruits to the full feeder, and examine which network at the hive—the waggle dance, trophallaxis, or antennation—best predicted when unemployed foragers would start arriving at the full feeder.[7]

A comparison of the importance of the waggle dance, trophallaxis, and antennation networks found that the dance network was the only one of the three that could predict when unemployed foragers would find work again by locating the productive food site. But in the honeybee world, sunset and inclement weather, such as rain and overcast skies, temporarily block out information about the position of the sun relative to the hive, a key bit of data transmitted during the waggle dance. When Matt and his colleagues designed an experiment in which honeybee workers fed from an established foraging site that was temporarily depleted but then relatively quickly replenished, mimicking bad weather and sunset, things played out differently. Now trophallaxis networks and antennation networks were the key to understanding the order in which foragers return to the replenished patch, with dance networks also playing a role, albeit a weaker one.

A wonderful sort of strategic redundancy is built into honeybee communication networks: if information about the sun relative to the hive is lacking, so that the dance network is shut down, the other networks are in place to pick up the slack, so that pollen and nectar continue to flow into the colony. No matter what the details of the problem, for honeybees, information about food for the hive is communicated along one or another social network.[8]

Animals communicate with one another about so many things: food, predators, mating preferences, dominance status, where to live, and more. Regardless of whether communication is honest or not, whether it is visual, auditory, tactile, chemical, or some combination of those, that information often travels along social networks—including the song networks of birds.

As a group, songbirds are the unrivaled champions of auditory communication in nonhumans. The production of song is the result of a wonderfully complex, and far from completely understood, combination of innate predispositions, trial-and-error learning, and social learning. So complex, in fact, that researchers use birdsong as an animal model for the evolution of human language. Over the years, birdsong has been studied by naturalists, animal behaviorists, evolutionary biologists, neurobiologists, comparative psychologists, and linguists. Dominique Potvin wanted to add another dimension to the study of birdsong in the silvereyes (*Zosterops lateralis*) that she studied all across Australia, and to do that she turned to both population-level thinking and social networks.[9]

In the chapter 6 discussion about the safety networks in place at Masai Mara National Reserve in Kenya, we saw that while nodes in a network most often denote individuals, they can also represent species. But nodes can also represent something else: populations. "I thought if you can look at how strong connections are between individuals in a population based on how many interactions they share," Potvin says, "why can't you do that with [the song] vocabulary that each *population* [of silvereyes] has and how much of that they share with each other?"

With their lovely olive plumage and distinctive white eye-ring, silvereyes are a speck of a bird, weighing about 14 grams and measuring less than 12 centimeters tip to tip. Males do the singing, and those songs are composed of syllables, each of which has slightly different acoustic properties in the 2–6 kilohertz frequency range. A male can produce sixty or so different syllables, but any given song is made up of between four and twenty syllables.

Potvin was interested in the role of song outside of the breeding season and focused on silvereye songs produced during their daily morning chorus. Never one to think small, for her dissertation she studied the songs produced in fourteen different populations—at seven locations, each of which included a rural and urban population—scattered across nearly a million square kilometers of Australia. One population at the National Botanical Garden was a short hop from the University of Melbourne, where Potvin was located; others were much farther away from her home base, including Brisbane to the northeast, Adelaide to the northwest, and Hobart, on the island of Tasmania, to the south. While silvereyes usually remain in their home population, Potvin and others know that occasionally silvereyes move back and forth as far as between Hobart to Brisbane, a flight of about 2,000 kilometers; the birds' slightly different plumage patterns, as well as different genetic signatures, help distinguish which area they are originally from.

Gathering data on songs from silvereyes from across those sites made for lots of time on the road. Potvin spent about two weeks at each of the locales intensely sampling genetics and song. In the more rural sites she camped out, but if she was at an urban site, she'd stay with a friend or at "a friend of a friend of a friend's place . . . using my own social networks." For each of the fourteen populations, Potvin devoted the first few days at a site to capturing male silvereyes, taking blood samples for genetic analysis, and placing a unique combination of colored leg bands on each bird. The color bands allowed her to be certain she wasn't recording the same individual day after day, but rather getting a broader sample.

Each morning Potvin got up early and headed out, so she could be with the birds when the morning chorus began. To get the data on the songs males were producing, she would peer through her binoculars and pick out a male and note its identity. Then she'd put on the headset that was attached to a "shotgun microphone" that allowed a single male's song, and nothing else, to be recorded. For months, day after day, site after site, population after population, Potvin repeated this process. For the most part, her memories of these road trips are fond ones, but what happened at one rural site in Brisbane makes her sad to this day. She'd caught and marked all the birds, and done some recordings, but didn't get all the data she wanted and decided she'd return when she finished with the urban site in Brisbane. Shortly before she left, Potvin was told by a local work crew that they were cutting a road through the area. Off she went to the urban site, gathered the data she needed, and two weeks later returned to the rural site. "Am I in the right place?" she asked herself. "I barely recognized it. I was devastated." She heard no songs in the now-barren landscape.

After all that time on the road, Potvin spent months analyzing the song data by using software that allowed her to measure the syllables and songs sung at all fourteen locations. One question she was interested in for her dissertation work was whether songs sung in urban populations were adapted to the soundscape of life in villages and cities (in that sense, the rural sites she studied acted as a sort of control). They did: urban silvereyes tweaked songs so that they were better heard over urban background noise. But it wasn't until a few years later, when she got a faculty position at the University of the Sunshine Coast in Sippy Downs, Australia, that Potvin began to think about using the massive amount of data that she had collected to build birdsong networks.[10]

That's where dragons come into the story.

One of Potvin's colleagues at the university was Celine Frere, whose many interests include animal social networks. In 2010, while Potvin was still working on her dissertation, Frere began a long-term study, still going strong today, of more than 350 eastern water dragons (*Intellagama lesueurii*) in a large park in Brisbane. Aside from the five months the dragons are in hibernation each year, Frere's team studies the social behaviors of the animals. Soon they were seeing evidence of friendships between certain dragons, which led Frere and her team to study their social networks. As Potvin and Frere got to know more about each other's research, Potvin told Frere about all the data she had on silvereyes song and that it might be ripe for a social network analysis. Frere agreed.[11]

For the network analysis, Potvin and her colleagues used data from some of the populations she had studied for her dissertation and supplemented that, working with a total of eleven mainland populations of silvereyes, as well as populations of silvereyes on Tasmania, Lord Howe Island, Chatham Island, and Norfolk Island, for a total of nineteen populations that became nodes in a social network. More specifically, the syllables sung by birds in a population, weighted by the relative occurrence of those syllables, became nodes in a song network, with the ties connecting the nodes representing similarity in the number and extent of use of syllable types between populations.[12]

Potvin and her team were interested in using network analysis to understand the connections, or lack thereof, across populations. One thing that mattered was genetic relatedness. When all those blood samples were analyzed and pairwise genetic relatedness across silvereye populations was calculated, Potvin's team found that the more related individuals in two populations were, the more similar were the songs sung in those populations in the network.

The silvereye team next turned to the question of whether the distance between two populations might also help explain the songs sung in their network of nineteen populations. To do that, they first calculated the centrality of each population, which measured how similar the repertoire of each population was to other populations. Next, they calculated the geographic distance between a population and its nearest neighbor to see whether centrality and nearest neighbor distance were positively correlated: Were populations that were central to the network also geographically close to their nearest neighbor? They weren't. But geography did matter in song networks in another way.

Populations at similar longitudes (north to south) were

more likely to have similar songs than populations at similar latitudes (east to west), regardless of how far apart they were; two populations at a similar longitude, but far from each other, were more likely to share songs than two populations at a similar latitude that were geographically close. “There are a few more mountain ranges as well as deserts and things like that, which are in the way of east-west movements,” Potvin says, and that should slow down the movement of information in her birdsong networks. But she thinks there may be even more to the latitude and song network story than just the hurdles posed by the Australian Alps, Blue Mountains, and the Tirari-Sturt Stony Desert, and she continues to probe into just what that might be.

None of this takes anything away from the sheer unadulterated beauty of birdsong. What it does is show that even one of the most wondrous of all forms of communication travels along social networks.

Back in the Budongo Forest in Uganda, the chimpanzees that Anna and Sam Roberts study were teaching them that communication networks are about more than dancing and singing: gestural communications are also embedded in a social network.

Using gestures produced by twelve of the chimpanzees Anna Roberts had studied during her PhD, Anna and Sam Roberts constructed communication networks in the Sonso chimpanzee community. For each of thirty-eight visual, twenty tactile, and fourteen auditory gestures, they created a proximity network—how close were chimpanzees to one another—as well as a behavioral network that examined the frequency at which gestures were displayed.

Results from their network analyses show the wonderfully complex social dynamics that we have come to expect in our nearest evolutionary kin. Preferred partners in a network—friends, who spend a lot of time near one another—exchanged many visual gestures with each other. Prior work in chimps had found that being in the presence of a friend was linked to lower heart rate. Anna and Sam Roberts think it is likely that visual signals are also associated with lower levels of arousal, allowing the gesturer to convey an emotional state to the recipient in a relatively stress-free manner.

The situation was different when gestures were exchanged between chimps who were not friends. In such interactions, network members tended to rely on auditory and tactile signals. Here, the Anna and Sam Roberts propose that auditory and tactile gestures are less ambiguous than visual gestures and may be especially important in reducing the uncertainty about the course of events that might follow from interacting with less familiar partners, a constant challenge in the fission-fusion society that exists in the Sonso community. This may be particularly important for tactile gestures, which are almost always associated with prosocial behavior, as these may allow chimpanzees to interact in larger social networks where friends would be relatively scarce compared to less famillіar acquaintances.[13]

From honeybees to silvereyes to chimpanzees, the animal world is awash with complex information being communicated along social networks. In the next chapter, we will up the ante on that complexity and delve into how the cultural transmission of new traits travels along social networks in everything from humpback whales and bottlenose dolphins to chimpanzees and great tit birds.

9. The Culture Network

He who receives an idea from me, receives instruction himself without lessening mine; as he who lights his taper at mine, receives light without darkening me. —Thomas Jefferson to Isaac McPherson, August 13, 1813

A core tenet of evolutionary biology is that for the process of natural selection to act on a behavioral trait, three things must be in place. First, there must be different behavioral variants in a population. For example, if the behavior is mate choice, then some animals might prefer colorful, risk-taking mates and others might prefer drabber colored, risk-averse mates. Without behavioral variation, there is nothing for natural selection to *select* on. But variation itself is not enough for natural selection to act: that variation has to map onto reproductive success, even indirectly. If behavioral variation does not translate into fitness differences, it does not lead to the evolutionary change. And, finally, behaviors have to be transmitted, with a reasonably high degree of fidelity, from generation to generation. Without a system of inheritance, any fitness differences associated with behavior in one generation are washed away in the next.

When Austrian monk Gregor Mendel's experiments on peas (among other things) were rediscovered in the early 1900s, it became clear that genes are one means of transmitting traits, including behavioral traits, across generations. Indeed, for the next fifty years or so, evolutionary biologists and animal behaviorists assumed genes were not only one way to transmit traits across generations—they were the *only* way. But in time, ethologists began to learn of another transmission system. In one study on the foraging behavior of Japanese macaques, primatologists Shunzo Kawamura and Masao Kawai found that behavior could be passed down across generations in nonhumans not only genetically but via *cultural transmission*. What's more, cultural transmission occurs not just between generations, from parent to offspring, but also within generations, from peer to peer.

To see how all this works, we turn briefly to Imo, a Japanese macaque monkey who lived on Koshima Islet, Japan, in the 1950s. Kawamura and Kawai and others studying Imo's troop of macaques threw sweet potatoes and wheat on the sandy beach so the monkeys would get used to the presence of humans. When Imo was a year old, she began to do something no macaque on the island had ever been seen doing: she washed her sweet potatoes in water before she ate them, dislodging all the sand and making sweet potatoes all the tastier. If Imo were the only one who did this, her innovation would have died with her. Instead, many of Imo's peers and relatives learned the skill of potato washing from Imo via cultural transmission: by watching her clean her potatoes and trying it themselves.

Imo's standing as a cultural icon among the macaques on Koshima Islet only grew stronger. When she was four, she found a better way to handle the wheat the humans put on the beach. Wheat and sand mixed together are edible, but not especially appetizing. Imo came up with a novel solution to that problem too. She tossed her wheat and sand mixture into the water. The sand sank and the wheat floated. This is a bigger deal than it sounds, because primates virtually never let go of food once they have it their hands. But to get the sand off, Imo had to release the wheat, and that's what she did. And as with the sweet potatoes, Imo's troop mates learned this new trick from her. More than sixty years later, long after Imo's death, macaques on Koshima Islet still wash their sweet potatoes and clean their wheat, and they, and we, have Imo to thank for it.[1]

Kawamura and Kawai didn't use social network theory to understand how Imo's foraging innovation spread through her group, for the simple reason that social network theory per se didn't yet exist. But many decades later when Janet Mann was studying culture and tool use in bottlenose dolphins in Australia, it did.

Tool use does not necessarily involve crafting wheels or gears—a definition that almost guarantees it will be seen only in humans—it simply means shaping an object to serve some function. New Caledonian crows (*Corvus moneduloides*) extract insects from under tree bark using tools constructed from twigs whose leaves have small sharp barbs running along their edges. Insects react and grab the inserted twig, and the crows pull out the twig tool and eat whatever is clamped on. Novice crow toolmakers begin by constructing simple tools, and as they mature, they start to build more complex ones. New Caledonian crows can even build tools from multiple parts, and remarkably they safeguard their best and favorite tools to reuse.[2]

Tools solve problems. For the bottlenose dolphins (*Tursiops aduncus*) in Shark Bay, Australia, the problem centers on poking and probing the sand with their snout to scare up fish like the spothead grubfish (*Parapercis clathrata*) from the bay bottom. It is anything but pleasant pounding your snout betwixt and between the rocks down there, so a small fraction of dolphins in Shark Bay, primarily females, pick up a basket sponge using their snout, wiggle that snout snuggly into the sponge, and use it to cushion the blow of probing the rocky bay bottom, effectively turning another organism into a tool. Shark Bay females take the choice of sponge seriously, often spending ten minutes finding just the right size and shape to fit their snout, and then swimming to their favorite hunting ground along channels between 8 and 14 meters deep. If a sponger does shake loose a bottom-dwelling fish hiding in the sand, they toss the sponge off and chase their prey. When they are done with that, if the sponge was a particularly good fit, they go back and retrieve it for further use.

Shark Bay, near Monkey Mia, sits on the west coast of Australia, about 850 kilometers northeast of Perth. The bay is huge, covering 13,000 square kilometers, with an average depth of about 9 meters, and is home to not only more than a thousand bottlenose dolphins, but also 16,000 dugongs (*Dugong dugon*), a marine mammal whose closest living relative is the manatee. Within its waters, it also has the world's largest seagrass banks, and the stromatolite fossils at Hamelin Pool Marine Nature Reserve in the bay date back almost 3 billion years. And, of course, Shark Bay is home to many sharks, some of whom feed on dolphins.

Since the Shark Bay Dolphin Project began in 1984, team members have gathered data on more than 1,500 bottlenose dolphins. Researchers have amassed a giant dolphin mugshot book identifying individual dolphins based not only on sex, but by using fin shape, as well as the shape of scars, often from shark bites, on the fin and body.[3]

Even before sponging behavior was discovered at Shark Bay, Janet Mann was there studying the dolphins. In 1988 she was a PhD student at the University of Michigan, working with animal behaviorist and primatologist Barbara Smuts. Smuts had become interested in the mother-calf interactions in the Shark Bay dolphins. One of Smuts's assistants on the dolphin project canceled at the last minute, and because Mann had done some undergraduate work on mother-infant interactions in baboons at Amboseli Park in Kenya, Smuts asked her to step in and help with the dolphin work, which Mann was more than happy to do.

Mann was part of the Shark Bay research team that first documented sponging behavior in 1995, and for almost thirty years, she and her colleagues have diligently been piecing together how this behavior has spread along social networks to a subset of dolphins in the bay. As in the literature on human culture, in animal behavior literature, one of the defining features of culture is that it differentiates groups from one another: "I was very much interested . . . in whether the dolphins had a concept of 'this is the way we do things and it's different from the ways others do it,'" Mann says. "Does that bind them together? Social network analysis lets you get at this with a mathematical method." Fortunately, Lisa Singh, a computational computer scientist at Mann's home institution, Georgetown, was already working with Mann to develop a custom database for the dolphins at Shark Bay. Singh also had experience working with social networks, and soon she, Mann, and their collaborators were looking at data on culture and sponging behavior from a network perspective.

When Mann and her colleagues began their social network analysis, they knew quite a bit about sponging behavior and spongers in Shark Bay. They were certain sponging worked as a foraging technique, not only because they could, if conditions were just right, see the dolphins sponging as they watched from their boat, but because they did some sponging themselves. They'd dive in an area that sponging dolphins liked and videotape the dolphins. Then, they'd do a second dive at the same spot, and while some divers videotaped, others probed the bottom wearing sponges on their hands. When they compared the videos, they were amazed by just how many bottom-dwelling fish were stirred up when human spongers shook up their otherwise safe world.

Mann knew spongers tended to be female. While the occasional male puts a sponge on and probes the bay bottom, for the most part males are more interested in forming coalitions and alliances that help them secure mating opportunities than they are in specialized tool use. Mann knew that except for a dependent calf who might be swimming beside her mother, when female spongers foraged, they did so alone; but dolphins who were not spongers tended to search for food in groups. She also that knew spongers—like Rub-a-Dub, Hubba, Grub, and Pub—always had a mother who was a sponger, and she hypothesized that young dolphins like Rub-a-Dub were copying—a type of cultural transmission—their mother's sponging behavior.

Cultural transmission from parent to offspring can be hard to nail down because it looks a lot like genetic transmission of behavior from parent to offspring. In both, *something* is passed down from a parent. In cultural transmission, that something is a skill, like how to forage using a sponge; in the latter case, it is genes associated with the behavior that are transmitted across generations. Behavioral ecologist Michael Krützen led a team of researchers, including Mann, who employed molecular genetic tools to rule out genetic transmission: sponging is indeed a culturally transmitted trait from mother to offspring.[4]

What Mann didn't know was whether female spongers were somehow networking with one another. Recall that sponging is a solitary affair, except for a female who sometimes has a dependent calf with her, which meant that spongers were *not* networking *while* they sponged. But sponging behavior takes up only a small part of a dolphin's day. Perhaps adult female spongers network with one another when they're not sponging? "Spongers [sometimes] carry their sponges and will join with other spongers with sponges on," Mann says. "They aren't foraging then, just hanging out." What's more, females who use a sponge once tend to specialize in sponging, which means that if you are sponger, even if another sponger isn't wearing a sponge when you see her, as long as you've seen her with a sponge on at some point, you can be fairly certain she is still a sponger.

To construct bottlenose association networks, Mann and her team looked for who was swimming near whom when dolphins were *not* foraging. Dolphins were considered to be associating if they were within about 10 meters from one another or were linked to each other through another individual using the same 10-meter rule. Mann used historical association patterns of 36 spongers and 105 non-spongers gleaned from almost 15,000 surveys that detailed where dolphins were between 1989 and 2010, what they were doing, and which dolphins they were swimming with.

What the social network analysis of these 141 dolphin networks revealed was that non-spongers were better connected than their sponger counterparts. Non-spongers had more associates and stronger bonds with those associates than did spongers. Non-spongers also had more friends and friends of friends. Part of the deeper connectedness of non-spongers in the network is due to their tendency to forage in groups: while foraging behavior was not part of the network analysis per se, non-spongers probably spend more time around others in anticipation of group foraging events.

The one major social network metric for which spongers

had higher values than non-spongers was the clustering coefficient, which, among other things, measures cliquishness. Non-spongers may have been more connected in the overall network, but spongers were forming cliques within that network, preferentially interacting with each other rather than non-spongers. But why? What were spongers getting from membership in a clique? If sponging is a solitary affair, why bother hanging around other foragers? Mann and her colleagues think at least part of the answer is that although a sponger always learns the skill of sponging from its mother, and only its mother, it may be that spongers can get another bit of important sponge-related information—namely, where the good sponges are located—from swimming around with others in their social network clique.[5]

Remarkably, dolphins are not alone in using sponges and in utilizing networking as a means for using them effectively. Chimpanzees also use sponges, but they network in a very different way than bottlenose dolphins. But before we compare sponging across species, let's peer in on the culture network of a cetacean relative of those bottleneck dolphins.

"I was one of those kids," Jenny Allen recalls, "that said I was going to grow up to work with whales, and everybody kind of went 'Yeah, OK, but, like, you're going to get a real job eventually.'" Eventually she did get a real job—working, as she predicted, with whales. As an undergraduate, Allen studied marine biology and landed an internship at the Whale Center of New England in Gloucester, Massachusetts. Soon she was spending her time on boats, studying humpback whales (*Megaptera novaeangliae*) in the Gulf of Maine's Stellwagen Bank National Marine Sanctuary, about 50 kilometers southeast of the Whale Center. When she graduated, Allen got a full-time position at the Whale Center and continued her work along Stellwagen Bank, watching the magnificent creatures she so loved.

Whale research can be tricky, as it often involves a deep tie with ecotourism and whale-watching expeditions. Allen had connections with whale-watching companies in the area, and she'd go out with them as an onboard naturalist. She'd narrate the adventure for the one or two hundred people on the boat, and, in return, she and her assistant were permitted to gather basic information on the whales, including their feeding behavior. They would also take photos, focusing on a whale's tail—the pattern on the tail is the equivalent of a whale fingerprint—and comparing it to the book full of tail shots they had with them.

Usually, the arrangement that Allen had with the captain of a boat worked well, but at times it could be a delicate balancing act. On a day when a ship wasn't encountering all that many whales, Allen might be ready to move on after fifteen or twenty minutes with a whale they did encounter. But she would often have to spend time trying to persuade the captain, whose job it was to maximize the time passengers on board could spend with the whales they paid good money to see, that he should indeed move on and hope they would encounter another whale. Unless it was just the right captain, in just the right mood, the ship stayed perched in front of the whale who was a sure thing.

Standing on deck, Allen was often transfixed by a strange behavior that seemed to be spreading across the population of the 1,300 whales she was studying. Whales in the Gulf of Maine (and elsewhere) had long been known to use a feeding technique called bubble-feeding, in which a whale dives 20 meters below the surface and gives off five or six blasts of breath around a school of prey fish. This has the effect of creating a bubble ring underneath and around the prey, and when the bubble ring reaches the surface, the whale crashes though the bubble ring, grabbing a meal as it does. But all of that wasn't enough for one humpback whale, who sometime in 1980 added a new twist to the bubble-feeding sequence. Right before diving to commence a bubble-feed to capture a school of fish, that whale slapped its tail on the water surface, in what has been dubbed lobtailed feeding. Soon, a few other whales were doing the same—though the number of slaps at the water surface varied from whale to whale—and by the time Allen was watching in the early 2000s, nearly 40% of the whales were using lobtailed feeding. It seemed to Allen that the lobtailing was spreading through whale networks via cultural transmission: naive whales who spent time with lob-tailers seemed to learn this new tradition by observation. Still, she knew that she couldn't be anything close to sure of that based solely on her observations.

Allen's interest in the role that social networks might play in the spread of lobtail feeding via culture began in 2010 when she entered the master's program at the University of St. Andrews in Scotland. There, she worked with Will Hoppitt, the network-modeling guru from the previous chapter, and with Luke Rendell, an expert on culture in animals. "Look, do you have any projects that would make a good master's thesis?" Allen asked Rendell and Hoppitt. "[Because] if not, I have access to this data set—maybe we [could use] that?" Allen had kept on good terms with the Whale Center, and the data set she was referencing was brimming over with information on lobtail feeding in humpbacks in the Gulf of Maine since 1980. Fortunately, Hoppitt was developing some new social network models and had not yet had the chance to apply them to a wild population or to a data set as massive as the one Allen had access to, and he was excited to do so. Rendell, who knew a great opportunity to study cultural transmission when he saw one, was keen on the idea as well.

Allen's thesis involved getting the nearly thirty years of data on lobtail feeding at Stellwagen Bank into a form that could be used in a social network analysis. That wasn't easy. Over the course of three decades, that data had been collected by dozens of different people, with varying degrees of expertise on foraging behavior in humpbacks. "The [observer] didn't necessarily say, 'This animal was lobtail feeding,'" she notes, "[but] because I could speak the language of that data collection . . . I [could] say, 'Oh, this sequence of behavior is lobtail feeding.'"

Allen searched through 73,790 time-tagged whale-sighting records, involving 653 individuals, each of whom had been seen at least twenty times. She coded a whale with a zero if, up to that point in time, it had never been seen lobtail foraging, or with a one, if it had been seen lobtail feeding at any point in the past. In addition, for records involving multiple individuals, Allen noted which whales were found together. Even after all that coding was complete, there were still technical hurdles. The data set was so large, that the computer model Hoppitt had built couldn't handle it all and crashed. But Hoppitt was up to the challenge, made the necessary adjustments, and soon the data and the model were syncing up.

If whales who were naive to lobtail foraging picked up this technique by watching lobtailers—that is, if lobtailing was spreading through the whale social network via cultural transmission—then, as a general rule, lobtailers should be seen associating with lobtailers. To see why, imagine a group of five whales who tend to spend time together. Suppose three of the whales are lobtailers and two aren't. If the two naive whales learn lobtail foraging from the others, then the next time that group of five whales is observed, and every time after that, they will all be lobtailers. The social network model that Hoppitt built was designed to ask if this was in fact what the nearly 74,000 records was showing.

What Hoppitt's model did was compare the probability that a naive whale was learning how to lobtail by observing lobtailers versus the probability that it somehow learned lobtail feeding on its own, by simple trial-and-error learning. To make that comparison, the model calculates a "social transmission effect," which captures how much interacting with knowledgeable individuals accelerates the pace at which a whale picks up the lobtailing behavior, relative to trial-and-error learning. The social transmission effect was huge and the results striking: so striking that Rendell didn't believe what he was seeing. "Luke basically was like, 'Oh, you've done something wrong,'" Allen says. "'You have to go back [and reanalyze].'" Allen, Rendell, and Hoppitt did check after check of the code and the data, looking for some sort of error or artifact, but the results held up. Still, Allen and her colleagues wondered whether there was some other, possibly simpler alternative than cultural transmission that could explain why lobtailers in the foraging network were associating with one another.

Allen knew that lobtail feeding most often occurs when whales are foraging on sand lance (*Ammodytes americanus*), a fish found in the Gulf of Maine. Maybe the reason that lobtailers were found together was not that lobtailing behavior spread throughout their foraging network by cultural transmission, but because the lobtailers were all drawn to the sand lance populations. To test for this possibility, Allen and her colleagues discarded all associations *after* a whale had been seen lobtail foraging for the first time. So, for example, imagine that whale 1 was spotted lobtail feeding on sand lance for the first time on December 31, 1999. Allen and her colleagues would know who whale 1 was swimming around with before that date. If they focused on that time period, they could test if whale 1 had been interacting with lobtailers before being a lobtailer itself. That meant they didn't need to worry that starting January 1, 2000, lobtailing whale 1 might be associating with other lobtailers because they all just cluster around sand lance. When they analyzed the data this way, they found that simply hanging around where the sand lance were did not explain why lobtail foraging had spread as it had. As Allen had long thought true but had no way of knowing for certain until they took social networks into account, information about lobtail feeding was spreading along the humpback whale foraging network via cultural transmission.[6]

It's nearly impossible to do anything experimental with humpback whales in the wild—and likely unethical as well. Not so for the great tits of Wytham Woods near Oxford. There, Lucy Aplin, later of Australian cockatoo social network fame (chapter 3), could teach a few birds new foraging innovations and test whether those innovations spread via cultural transmission to others in their social network.

From the start of her work, Aplin knew great tits were smart birds, and that cultural transmission played a role in how information spread through their societies. Back in the day, Brits had their milk delivered in bottles with colored wax tops. In 1949, long before Aplin was born, and years before Imo the macaque was spreading new foraging techniques on Koshima, ethologists James Fisher and Robert Hinde wrote:

> In 1921, birds described as tits were observed to prise open the wax board tops of milk bottles on the doorsteps in Swaythling, near Stoneham, Southampton, and drink the milk. This is the

first known record of an act which has now become a widespread habit in many parts of England and some parts of Wales, Scotland, and Ireland. The bottles are usually attacked within a few minutes of being left at the door. There are even several reports of parties of tits following the milkman's cart down the street and removing the tops from bottles in the cart whilst the milkman is delivering milk to the houses.

Prying off the milk tops to get at the contents was recorded most often in tits, including great tits (*Parus major*) and blue tits (*Parus caeruleus*). Fisher and Hinde had circulated a survey to 200 members of the British Ornithological Society regarding milk bottle opening, and from that they pieced together the history of the spread of this novel behavior over a large range in Great Britain. The data from the survey suggested that on occasion, a great tit accidentally stumbled upon the trick of how to get at milk inside one of the milk bottles in its neighborhood and that other birds learned this skill, at least in part, from watching the original thief. What's more, milk of different qualities had different-colored tops, and birds in an area where milk bottles were opened tended to prefer the same-colored tops, which strongly hints at cultural transmission. Those great tit milk thieves were not living in Wytham, but if great tits could figure out how to swipe the milk meant for a nice cup of tea, they seemed ideal for an experiment explicitly on cultural transmission and social networks.[7]

The great tit populations that Aplin worked with at Wytham are one of the best-studied populations of birds in the world, having been monitored by researchers from the Edward Grey Institute at Oxford for decades. The Wytham Tit Project began in 1947 when David Lack, a legend in the field of ecology, placed 100 nest boxes in the woods there. By 1960 the number of nest boxes had grown to 1,000, scattered all throughout the 385-hectare woods, known as "the laboratory with leaves."

Shortly before Aplin arrived, Ben Sheldon's team at Wytham had received a large grant, and part of the funds were used to add tags, each with an embedded microchip, on the great tits in Wytham. Rather than implant these PIT tags, as is often done, the Wytham group found a much less invasive method. All the great tits had leg bands that allowed researchers to know who's who, and the PIT tags, enclosed in a waterproof casing, were attached to those bands.

These PIT tags were critical for Aplin's work on networks and culture. What she—along with Damien Farine, Sheldon, and others—did was introduce a new foraging technique into a great tit population and test whether it spread through the social network by cultural transmission. To do this, they built a "puzzle box" that was stocked with mealworms, a great tit favorite. The mealworms could be accessed when a bird *either* slid a blue-colored door on the box from the left to the right *or* when it slid a red-colored door from the right to the left.

Once Aplin and her team had the puzzle boxes ready, two males from each of eight different tit populations at Wytham were captured and brought into the lab. The birds from two of these populations were trained as tutors to open the blue door (left to right), and the birds from three populations were trained as tutors to open the red door (right to left) to solve the puzzle box problem. Birds from the other three populations were not trained to solve the puzzle dilemma, and so served as controls.

After four days of training, Aplin released the sixteen males in the lab back into the populations from which they had been captured, but not before she and her colleagues placed three puzzle boxes into the home range of each population. Each puzzle box had been fitted with a radio-frequency identification (RFID) antenna so that each time a bird arrived at a puzzle box, its PIT tag ID was logged, as was data on whether it moved a blue door left to right, a red door right to left, or flew away failing to solve the puzzle.

While all of this was going on, Aplin's team and others on the project also gathered data on groups of tits while they ate at a separate set of feeders full of sunflower seeds that were *always* available to all the birds. The sunflower seeds at these feeders were freely available—no puzzles needed to be solved there—and the interactions between birds at these seed feeders were used to construct tit association networks.

All of this had Aplin and her colleagues on the move on their bicycles and motorbikes. Data had to be downloaded from the puzzle boxes, mealworms at those boxes had to replenished (or replaced if rotten), the batteries that charged the RFID antennae had to be swapped out when depleted, and seeds had to be constantly added to the sunflower seed feeding stations. "It was a lot of carrying heavy loads and walking after dark in the woods," Aplin says, "but it was great, great fun." Great fun and remarkably informative.

The first clue that the solution to the puzzle box was moving along social networks via cultural transmission was that birds from the control populations (where the birds had been captured, but not trained) solved the puzzle box problem far less often than birds from populations with tutors. But there is way more to it than that. After the twenty days that puzzle boxes remained out in the woods, an average of 75% of the birds in populations with tutors were solving the puzzle using the same technique the tutors in that population used. The puzzle boxes were opened more than 57,000 times, and from the original ten tutors in the five populations that had tutors, more than 400 birds were now opening the boxes as their tutors had.

Aplin and her colleagues then used the social network data from the seed feeders to see who associated with whom, and then mapped the puzzle box data on to that. What they found was that the more deeply tied an individual was to a tutor in their association network at sunflower feeders, the more likely it was to have learned to open the puzzle box in the same manner as its tutor. Indeed, for every additional unit of time a bird spent associating with a tutor in the association network, it was twelve times more likely to be able to solve the puzzle box.

Clearly the information on how to solve the puzzle was being transmitted culturally along tit association networks, but Aplin wanted to know more. Just how important was this networked culture to the birds? The answer, she discovered, was very important. If after learning to open the puzzle box as its tutor had, a bird stumbled upon the other way to open the box, it generally stuck with what it had learned from its tutor. And culturally transmitted foraging traditions had another kind of staying power. Almost a year later after the puzzle boxes had been removed, Aplin put them out again for five days in the home area of one population that originally had blue door tutors and in one population that originally had red door tutors. Despite the fact that no *new* tutors were introduced at that point, and that there had been significant population turnover, birds in each population quickly began opening the puzzle boxes using the technique that tutors had introduced a year back. Some of these birds had learned how to open puzzle boxes from tutors the prior year and just rekindled a previously culturally acquired behavior, but other birds were learning for the first time from either the original

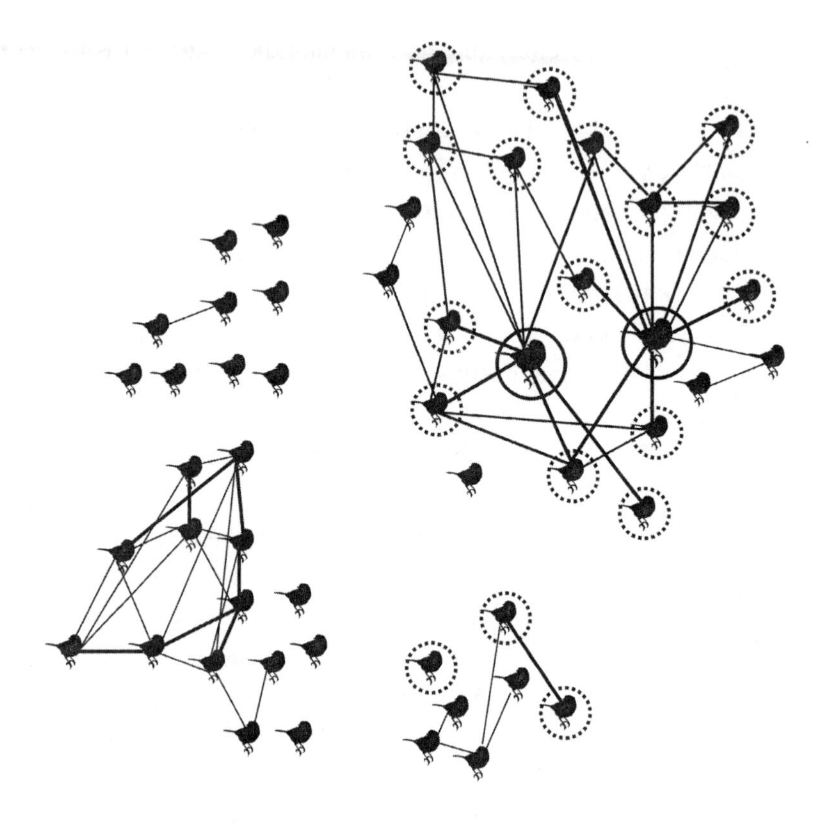

Schematic of foraging network of great tit birds in Wytham Woods, near Oxford. Two individuals (in closed circles) were trained to access food in a novel way and released back into their population. Broken circles show individuals who acquired the new foraging technique. Birds not in circles failed to acquire the new foraging technique. The new technique diffused through the subsection of the population that interacted most with tutors. Bird image with permission from istock.com. Networks loosely based on L. M. Aplin, D. R. Farine, J. Morand-Ferron, A. Cockburn, A. Thornton, and B. C. Sheldon, "Experimentally Induced Innovations Lead to Persistent Culture via Conformity in Wild Birds," *Nature* 518 (2015): 538–41. Great tit bird image with permission from istock.com.

tutors (who were still present) or from birds that had rekindled what they had learned from tutors earlier.[8]

Returning to tool use, cultural transmission, and social networks, think back to the Sonso community of chimpanzees in Uganda's Budongo Forest discussed in the last chapter. At 9:05 a.m. on November 14, 2011, an innovation was born to chimpanzees in that community when a dominant male in that troop fashioned a new tool. And though she did not realize it at the time, it was also a special moment for primatologist Cat Hobaiter, who was right there watching. Hobaiter wasn't the only one watching, though: other members of the Sonso troop were as well, and for the next five days a new method of making tools spread via cultural transmission along the Sonso social foraging network.

It wasn't tools, culture, or social networks in chimpanzees that first brought Hobaiter to Budongo Forest for her PhD research. Instead, it was the same thing that brought Sam and Anna Roberts to Budongo: gestural communications. Hobaiter took a gestalt perspective on those gestures: "I think understanding a system of communication requires that I understand everything else about the chimps' world," she says. "It's important to me to understand what it means to be a chimp; otherwise, why else would I care what their gestures mean?"

Understanding everything meant getting up early, grabbing her video camera, and following as many chimpanzees as she could that day, indeed for every day of her ten-month stints at Budongo. Fortunately, the field camp, where Hobaiter had a small house, was right in the middle of the Sonso chimpanzees' home range, and she could almost roll out of bed and start taking data. When there was data to take, that is.

"Ninety percent of what I do," Hobaiter says, "is wait around for the chimps to do something, usually for them to come down out of a tree where they are sitting in the beautiful, warm sunshine and I've been sitting in the mud on the dark, wet forest floor, covered with bugs."

On that fateful day of November 14, Hobaiter was following a group of the Sonso chimpanzees when they came upon a new watering hole at the roots of a tree. Watering holes are a dime a dozen in the rainforest, especially during the wet season, but for some reason, the chimpanzees really liked this watering hole. "They just went crazy for it," Hobaiter says. "It was gesture heaven. . . . Chimps telling one another to get lost and get out of here, or affiliate with me, or can I come and approach you." For six straight days she followed this group, videotaping everything they did.

Hobaiter noticed that Nick, one of the dominant males in the troop, was taking clumps of moss and molding them together into a sponge and sucking water from it. Although she had seen chimpanzees fashion sponges from leaves, Hobaiter had never seen a chimpanzee at Budongo fashioning a sponge from moss. But with her focus centered on gestural communication, she didn't make much of that, nor of the fact that others in the troop started doing the same thing. At the end of those six days, Hobaiter watched the videotapes back at camp, coded them for gestures, and that was that.

Months later, Hobaiter was out in the forest with her colleague Thibaud Gruber, a Swiss animal behaviorist. They had plenty of time to chat, and Hobaiter knew that Gruber not only studied communication, but also had an interest in tool use and cultural evolution. "[The chimpanzees] had been doing this kind of weird sponging behavior, using moss," she told Gruber, adding that she had six days of videotapes to prove it. "We were way down in the swamp," Hobaiter says, "and

Thibaud literally marched me back to camp to see the tapes." After they did, focusing this time on moss sponges, not gestural communication, they knew they were on to something. It looked like the use of moss sponges was spreading, via cultural transmission, through a subset of the Sonso population. But unlike the great tits, humpback whales, and bottlenose dolphins, where the spread of information was inferred by association patterns, the videotapes Hobaiter had held the potential for tracing the precise spread of this new tool technologically from the moment Nick first created a moss sponge.

When Hobaiter returned to her home base at the University of St. Andrews in Scotland, she recruited Will Hoppitt, who was at St. Andrews at the time, to build a new network model to see just what was happening with moss sponge use in Budongo. "I had this data set entirely by accident," Hobaiter says. "Thibaud had the research question for a long time, and Will had wanted to model something like this . . . a lovely coming together of all of our kinds of interests."

As Hobaiter and her colleagues went through the videotapes, they kept a log of the identity of every chimpanzee at that special watering hole and what they were or were not doing. After Nick first constructed a moss sponge, Hobaiter noted audience members present when Nick (and then others) made moss sponges. The audience bar was set high: a chimp was only considered a member of an audience if it was within 1 meter of an individual who was actively making a moss sponge and was either staring at the toolmaker, had shifted its eye gaze to the toolmaker, or had moved its head in a way that tracked the toolmaker's movements.

Nambi, a dominant female in the waterhole network, was the first to learn to fashion a moss sponge from watching Nick. From that point forward, seven of the other thirty chimps in that watering hole network began crafting moss sponges. Six of those seven chimps picked up that skill only *after* being in the audience of a sponge-maker. The audience effect was dramatic. Every time a non-sponge-maker was an audience member, its chances of learning to use moss as a tool increased fifteen-fold: if a chimp who did not display sponge-making was an audience member twice, its chances of picking up the skill increased thirty-fold.

It all made sense. All except for one thing. Kwara, a dominant female in the network at the watering hole, had never been in the audience of a sponge-maker, and yet there she was fashioning moss sponges. Cultural transmission along the chimp social network explained the use of moss sponges—except for Kwara. Hobaiter went back over the tapes to make sure they hadn't missed Kwara being in the audience at some point: they hadn't, but that second run-through of tapes taught Hobaiter that culture, and how it spreads new behaviors along social networks, was even more complicated than she had first thought.

It turns out that just minutes before Kwara made her first moss sponge, she had picked up an old moss sponge from the forest floor and used it herself. She hadn't seen anyone fashion a tool, she just found one lying there and gave it a try. "[Cultural] transmission in chimps," Hobaiter realized, "is not just about what they get to see, but also that they get to interact with the product of others' behavior." That very product itself—in this case a discarded moss sponge—plays an indirect role in network dynamics. Kwara was the fourth chimp to make a moss sponge: Janie, the fifth to pick up the skills of sponge-maker, learned by sitting in the audience when Kwara displayed her new toolmaking talent.[9]

If we step back and compare dolphins and chimpanzees, we see just how complicated and interesting the relationship between tool use, cultural transmission, and social networks can be. In the case of the chimpanzees, a network was in place at the watering hole and information about tool use spread along that network via cultural transmission. In the case of dolphins, cultural transmission of tool use was much more limited, involving just mothers and their offspring, and tool use was a solitary affair. Tool use did not move along large established networks. Instead, tool users actively chose to network with others who used those same tools.

Whether it's great tits in the Wytham Woods, humpback whales in Maine, bottlenose dolphins in Shark Bay, or chimpanzees in Uganda's Budongo Forest, cultural transmission plays a role in the dynamics of social networks. But as animal behaviorists and disease ecologists have come to learn, the interconnectedness that allows information—including culturally transmitted information—to move along social networks, also allows microbes (some beneficial to their hosts, others parasitic) to move from network member to network member. In the next chapter, we'll explore how this happens, and the implications of having animal social networks navigated by both welcomed and unwelcomed guests.

10. The Health Network

Every one of us is a zoo in our own right—a colony enclosed within a single body. A multi-species collective. An entire world.
—Ed Yong, *I Contain Multitudes* (2016)

Animal behaviorists, often working with geneticists and epidemiologists, have taken two different approaches to studying how animal social networks act as highways along which microorganisms can navigate from host to host. One is to focus on a specific microbe and examine how it navigates a network, and what that means for network dynamics in the animals whose networks are tapped into from below. A different approach uses sophisticated molecular genetic techniques to sequence as many bits as possible from a host's microbiome—all the microbial species that call that host a home, or at least a temporary home—and then explores the relationship between animal social networks and the microbiome members who hitchhike through those social networks. Elizabeth Archie and her colleagues have taken the latter approach in studying the social networks of savannah baboons (*Papio cynocephalus*) that are part of the Amboseli Baboon Research Project (ABRP).

The ABRP field site sits in the southwestern corner of Amboseli National Park. Under what Archie describes as "a huge sky," the baboons walk about in this short grass, savannah habitat that is, for the most part, flat and dusty, as well as salty. Here and there are the occasional acacia trees and fever trees. Mount Kilimanjaro in neighboring Tanzania, only about 50 kilometers south, dominates the horizon. As the baboons or humans traverse the ABRP site, they may see a herd of elephants, a pride of lions, a clan of hyenas, various species of gazelles, and perhaps some zebras. If they look up, they might be treated to superb starlings, ring-necked doves, or blacksmith plovers flying by.

The birth of the ABRP came in 1963, when ethologists Jeanne Altmann and Stuart Altmann were exploring East Africa for the perfect place to study baboons. When they visited the Maasai Amboseli Game Reserve—today's Amboseli National Park—they knew they had found what they were looking for. After spending thirteen months studying baboons there, they began planning something much more long term. They returned in the late 1960s, published *Baboon Ecology: African Field Research*, now considered a classic in primatology circles, and in 1971, together with a team at the University of Chicago, launched the Amboseli Baboon Research Project. In the course of the fifty-plus-year project, the ABRP has assembled an unparalleled compendium of data on some 1,300 baboons.[1]

Archie did her PhD work on elephants at Amboseli National Park, and when she was offered a postdoctoral position to help run the ABRP by Susan Alberts, already one of the directors of the ABRP, she jumped at the chance. Her route to studying the microbiome of Amboseli baboons using social network tools was just as circuitous as that which brought her to the baboons in the first place. In another postdoctoral position she had before becoming a professor, she spent time at the Smithsonian National Zoo. While there, she had the opportunity to interact with lots of people who were studying disease ecology. Before that she hadn't thought all that much about infectious disease and social behavior, but based on those interactions, she began thinking of animal social networks as akin to maps that track the spread of microbes.

At that time, Archie was mostly thinking about pathogens and ended up doing some work on disease ecology in elk and bison at the National Bison Range in Montana, "picking up poo from elk and bison and bighorn sheep and then extracting parasites out of them" to study the parasites moving among hosts. Soon she began to think she was looking at only one side of the microbial coin, the side with all the microbes that caused disease, but the microbiome, she knew, also had microbes that were beneficial to their hosts. Archie thought that she could combine her interest in social behavior and social networks with the use of genetic techniques to look at a whole community of microbes, some beneficial, some harmful. And the baboons at Amboseli—where grooming networks were part and parcel of everyday life, and there were plenty of feces to collect for microbiome analyses—were perfect for doing just that.[2]

When Archie arrives at the ABRP base camp, usually for three multiweek stints each year, she's immediately among friends, almost all of whom are Kenyans. "Our [Kenyan] data collectors," she says with a smile, "have been with us for decades." Both when Archie and her colleagues are at camp, and when they are not, the data collectors, working in pairs and taken to the baboons by a driver, census troops and gather information on all things baboon.

The camp covers about a hectare and is surrounded by a solar-powered electric fence, "because if you leave the camp," Archie says, "there's lions and elephants and buffalo and stuff you could walk into and would probably kill you." There are a number of camps around the area, each with its own solar-powered fence, but those are often down: the animals, including the elephants, are constantly looking for spoils and testing fences, to see if a breach is possible.

Archie gets up about 5 a.m., chugs some coffee, and eats a small breakfast before helping people pack up the car with a radio antennae, notebooks (and electronic tablets, more recently) for collecting data, and then off they go. The drive from camp to what Archie calls "baboonland" takes about forty-five minutes. Every baboon in baboonland has its own story to tell. Flank was one of Archie's favorite animals: "She was like the Velveteen Rabbit of baboons . . . patchy coat with tufts of fur sticking out everywhere and bald patches."

Each troop has its preferred sleeping roosts, so Archie and her team can usually guess where the troops are, but on occasion they'll surprise her and sleep somewhere new. No matter, as each of the five troops that are tracked in the ABRP has a few individuals that are collared with radio transmitters, and so once Archie and her team get to baboonland, they break out the radio antennae to find out exactly where the groups are that morning.

Once they find a troop, Archie and one of the two assistants she has with her get out of the car and approach the baboons: on a good day they can observe troop members from about 5 meters away. Then they perch themselves with binoculars and a notebook or tablet, and collect data. They know every baboon in a troop, and for the first hour or so, they just census the group. Then they start cycling through the group, watching one baboon for ten minutes, another baboon for ten minutes, and so on. As they do this, they are always on the lookout for a chance to swoop in to grab a fecal sample from a baboon who is gracious enough to make a donation for the microbiome analysis. When the vials eventually make it back to the United States, DNA is extracted from each, and a large molecular genetic assay, called a shotgun analysis, is run to identify as many microbial species as possible in a sample.

For the microbiome/social network study, Archie collected data from individuals in two troops called Mica's group and Viola's group, who have home ranges adjacent to one another. Across all baboons in these two groups, 925 microbial (bacterial and archaeal) species were identified in the fecal samples. The single strongest predictor of what a baboon's microbiome contained was whether it was a member of Mica's or Viola's group.

That suggested that microbes were moving along some sort of social network in Mica's group and Viola's group, but group membership might create microbiome similarity for any number of reasons not network related. Group members share the same watering hole and food sources, and so it might be that diet differences between Mica's and Viola's groups explain differences in the microbiome. That seemed unlikely given that the groups lived on adjacent home ranges, but Archie and her team checked to be sure, and they found that diet was very similar in both groups. So that wasn't it. Another possibility was that because genetic relatives often have similar microbiomes, and baboon groups are kin-based, kinship explained what they found. It didn't. Nor was it the case that groups differed in sex ratio or age structure, each of which could, in principle, create similar microbiomes within groups.

All of this strongly suggested that social networks within baboon groups might help explain microbiome differences between groups and microbiome homogeneity within groups. All the data Archie and others in ABRP had collected on grooming networks was perfect for addressing that question. What Archie and her colleagues found was that in both Mica's and Viola's groups, grooming partners who were more tightly linked to one another in the grooming network had more similar microbiomes.

But how does all this work? From the perspective of microbes, what is it about baboon grooming networks that allows them to leverage those networks to hop to a new host? Archie thinks it might be because same-sex grooming involves lots of hand-to-mouth contact, and grooming between males and females that are sexually receptive involves quite a bit of contact in the ano-genital region, making fecal-oral transfer a possibility. Such intimate contact between hosts may be particularly important to anaerobic microbes that can't survive long in the oxygen-filled environment outside of their hosts. Indeed, when Archie looked at those microbes in the guts of baboons who were most tightly tied to grooming networks, she found that species that can't survive long in oxygen were far more common than expected by chance.[3]

Microbes, particularly those that fare poorly outside the baboons' guts, are clearly using the grooming networks of the monkeys to move from host to host. What Archie and her team would really love to know now is what proportion of those 925 species of microbes are detrimental versus beneficial to the Amboseli baboons. The problem is that while the genetic shotgunning approaches they use are wonderful for identifying what is present, by themselves they don't tell you how microbes affect their hosts. For that, those microbial genomes need to be compared to other microbial genomes where the functions of key genes of interest have been determined, so that inferences can be made about whether they are likely beneficial or detrimental to the baboons.

While microbiologists and evolutionary biologists need to do a lot more work on the function of microbial genes before a complete picture emerges, Archie and her team did find some evidence for both beneficial and detrimental microbes cruising along the baboon grooming networks. For example, one microbe, *Campylobacter ureolyticus*, which causes gastrointestinal problems, resides in the baboon microbiome and moves from baboon to baboon along grooming networks (and perhaps in other ways as well). On the other hand, some members of the microbiome community of the baboons are known to have probiotic effects in humans, and so perhaps in baboons, by inhibiting pathogens and aiding in the production of vitamins. The good and the bad from microbe world both move along baboon grooming networks.

About 1,600 kilometers southeast of Amboseli, in the Kirindy Mitea National Park, on the coast of the evolutionary wonderland known as Madagascar, behavioral ecologist Amanda Perofsky has taken a similar approach to studying the social networks and microbiomes of a lemur called the Verreaux's sifaka (*Propithecus verreauxi*).

Kirindy Mitea National Park covers 6,000 square kilometers in southwestern Madagascar, in the Kingdom of the Sakalava, near the Mozambique Channel. A UNESCO Biosphere Reserve, the park is a unique mixture of desert, dry spiny forest and coastal mangrove ecosystems, and home to eight to ten (depending on which taxonomist you talk to) species of sifaka, including the endangered Verreaux's sifaka that Perofsky studies. In 2006 biological anthropologist Rebecca Lewis began a study of the ecology, behavior, genetics, and health of the Verreaux's sifakas in Kirindy, and that same year, she also established the Ankoatsifaka Research Station in the dry forest area of the park to facilitate that research. The station has a bare-bones kitchen that runs on solar power, tents for visiting researchers like Perofsky, and a few bungalows for the full-time Malagasy staff who keep the station running. That staff, which includes Andriamampiandrisoa Maximain and Nihagnandrainy Francisco, do more than maintain the station—they play an integral role in gathering data on sifaka foraging and social interactions.

Studying disease ecology had been a passion of Perofsky's ever since she took a course on the topic as an undergraduate at the University of Georgia. In time she got in touch with the sifaka research team at the University of Texas at Austin, headed by Lewis, who along with computational biologist, Lauren Meyers, served as her main PhD mentors. The initial plan was for Perofsky to study disease ecology in the sifaka, but for logistic reasons associated with the remoteness of the field site, her focus shifted to the sifaka microbiome and its ties to social network dynamics.

When Perofsky arrived at Kirindy, the Verreaux's sifakas, with their white-furred bodies and patchwork white and black faces, were leaping from tree to tree and, and when on terra firma, appearing to bounce around using an unusual and remarkable mode of locomotion called vertical clinging and leaping. Wonderful as that was to watch, Perofsky was there to study their microbiome and then to tie it to the behavioral work that others on the sifaka team were gathering.

Sifakas, which weigh a mere 3–4 kilograms, groom one another and, like in baboons, that often involves contact near the ano-genital region. Both males and females scent-mark their territories, often by rubbing their ano-genital region against trees. Individuals often scent-mark over the marks of others in their group, and that, combined with ano-genital grooming, facilitates microbe transmission within a group. And then there is the sifaka diet. Sifakas are folivores, drawing nutriments primarily from leaves, and their intestines are specialized for breaking down those leaves by employing bacteria in their microbiome that help in the fermentation process.

Perofsky and her colleagues collected data on forty-seven

sifakas who lived in seven groups, which varied in size from four to nine lemurs each. Others, primarily Lewis and the Malagasy team at the research station, gathered behavioral data on grooming, scent-marking, feeding, and dominance. To figure out who the denizens of the sifaka microbiome were, what Perofsky and Elvis Rakatomalala, a Malagasy assistant on the project, did was what everyone who studies microbiomes in the field does: they collected fecal samples immediately after the sifakas were kind enough, as it were, to provide them. It was not easy work, and if Lewis and the sifaka team had not cut a series of marked trails through the dense, spine-filled forest near the station years earlier, it would have been nearly impossible.

Perofsky would rise early each day so that she and Rakatomalala could reach the group they were focusing on that morning. All the lemurs have a collar for individual identification, and one female in the group has a radio transmitter on her collar, so Perofsky knew the location of the sleep tree for that group. She needed to be there before the sifakas woke up, so she'd be present to collect their first fecal deposit of the day, because once the lemurs start moving in all directions, it becomes hard to track all group members simultaneously. The samples Perofsky and Rakatomalala collected would be eventually transferred back to the University of Texas at Austin, where they were frozen until all the genetic material in them could be sequenced.

Perofsky was at the Ankoatsifaka Research Station for two different multi-month stays, and each time when she headed back, getting those critical fecal samples out of the country proved challenging. For one thing, she had to set up export permits, but to do that, she needed cell phone reception, which, to put it generously, was spotty. "Someone had thrown a rope over a very high tree branch and attached a little pouch to it," Perofsky tells it. "You'd put your phone in the pouch and pull on the rope, so you could get the phone up in the air to receive messages, and then pulled it back down so you could read them." But she also needed to send messages. For that, she'd walk about 1.5 kilometers from the field station, where one of the people at the station had attached a ladder to a tree. Perofsky would climb up and wave her cell phone and hope for a signal to get a message. Even after calls eventually helped secure export permits, the customs officers at the airport could never quite understand why anyone was taking two large trunks of preserved poop out of the country, which just added another layer of stress to the journey back home.

Perofsky discovered that the sifaka microbiome largely houses previously undocumented microbes: undocumented primarily because so little is known about lemurs, who are only distantly related to humans, chimpanzees, and baboons, from whom most microbiome data has been gathered. That said, two of the groups of microbes found most often in the sifaka microbiome have been looked at in other primates and are known to maintain gastrointestinal health, boost the immune system, and fend off infection.

As it was with the baboons at Amboseli, Perofsky and her team found group membership in sifakas had a powerful effect on microbiome similarity. Members of the same group were far more likely to have similar microbiomes than members of different groups. When Perofsky and her colleagues looked at what was going on within groups, there was a nice, clean relationship between the grooming network in a group and microbiome similarity between network members. Not every sifaka in a group grooms every other sifaka, but pick any two lemurs in a group, and the fewer the number of links it takes to get from one to the other in the grooming network, the more similar their microbiomes. What that translates into is that if sifaka 1 groomed sifaka 2, they had relatively similar microbiomes. But friends of friends also mattered. If sifaka 1 didn't groom sifaka 2, but groomed sifaka 3, who groomed sifaka 2, then sifaka 1 and 2 had similar microbiomes, though not as similar as bosom buddies who groomed each other.

What really jumped out to Perofsky and her team about what was happening within groups was that the higher a sifaka's centrality in the grooming network, and the more often it scent-marked, the more diverse the microbiome of that individual. And microbiome diversity matters when it comes to health, as research in other primates has shown that individuals with diverse microbiomes tend to be particularly healthy: centrality in a grooming network not only maintains social bonds, but it also fosters good health.[4]

Over in Woodchester Park, about 160 kilometers northwest of London, work on social networks and microbes isn't about the vast array of species that make up the microbiome; it's about a very specific microbe, because the resident badgers (*Meles meles*) there have a disease problem. What's more, the resident Welsh black cattle who graze in pastures in the park also have a disease problem. And it is the same problem: a nasty pathogen called *Mycobacterium bovis*, which in cattle causes bovine tuberculosis, a disease that costs the UK government £100 million annually to try to control.

Both cow-to-cow and badger-to-badger transmission of *M. bovis* have been well-documented, and there's lots of evidence that badgers spread *M. bovis* to cattle: badgers and cattle carry the same molecular variant of *M. bovis*, and outbreaks of bovine tuberculosis are more likely when badger populations reside near cattle pastures. What's more, culling the badger population decreases incidents of bovine tuberculosis, though it also causes some havoc when badgers try to flee the culling zone, as those who do escape often cause an increase in bovine tuberculosis in nearby cattle herds.

The catch is this: badgers and cattle almost never interact with one another. One study using camera traps found that in more than 64,000 hours of tapes, cattle and badgers were *never* seen in close proximity. How then does *M. bovis* get from badgers to cattle? Animal behaviorist Matthew Silk had an idea. For his postdoctoral work, he and others, including epidemiologists, thought that badger-to-cow transmission might take place, indirectly, via badger latrines, which are communal defecation and urination pits that sometimes demarcate group territories. To test that idea, Silk used a new type of social network analysis that he helped develop. Dubbed multilayer social network analysis, it examines interactions in multiple networks simultaneously, and Silk thought this approach might help sort out whether badger latrines serve as hubs linking badger social networks and cattle social networks.

Badgers live in about two dozen groups across a 7-square-kilometer section of Woodchester Park and have been monitored continuously since 1976, with data on more than 3,000 individuals gathered over the decades. The study area in the park is nestled in a wooded valley interspersed with cattle pastures where herds of Welsh black cattle graze. Badgers are trapped (and then released) periodically, and information on size, infection status, group membership, and more is collected.

The key to Silk mapping out the social networks in the badgers and the cattle was that forty-two badgers and twenty-five cattle were collared with proximity sensors like those discussed earlier in vampire bat networks. In the case of the badgers, Silk needed to make sure those proximity sensors had been set correctly at the factory, so that they were indeed talking to each other once badgers were about a meter apart, as he had requested when he bought them. This meant a bit of rather unusual calibrating that involved placing the proximity sensors on large empty saline bottles meant to mimic badgers, and then lying in the grass on the campus at the University of Exeter and moving those saline-bottle/pseudo-badgers around to see when they in fact started talking to each other. It was quite the scene. "We had people who thought we were incredibly strange," Silk says, smiling, "and others who were incredibly curious and would come up and ask questions." No one caused any trouble, though, as Silk explains, "at the peak . . . there were five or six of us doing this and we were probably an intimidating group all lying in the grass moving these bottles around."

When the calibrating was all done, the badger proximity sensors were set to talk to one another when two badgers were between half a meter and a meter apart: the cattle proximity sensors were set to turn on when cows were within 1.5–1.9 meters from one another. Badger and cattle proximity sensors also talked to one another, so that any direct contact between species could be logged, though based on those 64,000 hours of videos others had taped, Silk and his colleagues had little reason to think there would be much of that. Contact information was stored and could be downloaded from collars that housed the proximity sensors, unless a badger managed to remove the collar, which happened on occasion. Not only did the study involve proximity sensors on two species, but Silk and his colleagues placed base stations at nineteen badger latrines. Each base station acted as a sort of giant collar for the latrine, as it talked with the proximity sensors on both the cattle and the badgers, so the researchers knew which badger or which cow was near which latrine and when. What's more, the base stations could be used to build an indirect network of sorts between the latrines themselves using the movement of badgers and cattle between them.[5]

Once all the data had been downloaded from the collars on the badgers, cattle, and the base stations at the latrines, Silk used multilayer social network analysis to piece together what was happening. This sort of analysis can get very complicated, very quickly, particularly when looking at three networks (badgers, cattle, and latrines) simultaneously, but a number of findings rose to the surface.

Within the networks formed in each badger group, female badgers were more deeply connected (as measured by centrality scores) to others in the network than were males. Whether badgers were infected with *M. bovis*, though, had no effect on any network measure, suggesting that infected badgers were as socially connected as their uninfected counterparts, which is not good news in terms of disease transmission within badger groups. Proximity sensors detected no direct interaction between badgers in different groups/networks, but those sensors also showed that the latrine network indirectly linked badgers in different groups, and so served as locales where *M. bovis* is transmitted *between* badger networks.

Consistent with prior work, the proximity sensors recorded not a single direct interaction between badgers and cattle: they were never close enough to make their proximity sensors talk to one another. But, again, the latrine network was the key, as it facilitated the transmission of *M. bovis* between badgers and cattle. In general, cattle tend to avoid areas with badger latrines, but if the quality of a pasture is good enough, they'll graze there, even if latrines are present. What Silk and his colleagues found was that a small number of latrines in the middle of the pasture were visited by many badgers and cattle and had high centrality scores in the latrine network, serving as indirect hubs that link badger networks and cattle networks, allowing for the transmission of *M. bovis* between species.[6]

On the other side of the planet from Woodchester Park, in Tasmania, dangerous cells are also moving along animal social networks, this time the networks of the Tasmanian devil (*Sarcophilus harrisii*). But those dangerous cells are not quite microorganisms and not quite foreign cells, but instead a cancer that travels along, and simultaneously dramatically impacts, the biotic highways that Tasmanian devil social networks create.

Most cancers can't be transmitted between individuals, but a few, including Tasmanian devil facial tumor disease (DFTD), and canine transmissible venereal tumor (CTVT), are exceptions to the rule. DFTD is a rogue lineage of devil cells that proliferate rapidly. It is much more dangerous than CTVT, almost always killing a devil within a year of infection, and as a result, many populations of devils are in severe decline in Tasmania, with the IUCN listing Tasmanian devils as "endangered." Because devils are seen as something of a national icon in Tasmania, the devastation caused by DFTD is considered a conservation emergency: the Department of Natural Resources and Environment in Tasmania has a public outreach program called Save the Tasmanian Devil to inform Australians of the precarious state of the animals. But to save the devils requires an understanding not just of DFTD per se, but also Tasmanian devil social networks, as the cancer is spread when individuals bite each other, usually on the head and face. The social network side is where animal behaviorist David Hamilton comes into the story.

After Hamilton graduated from the University of Edinburgh, he spent a few years as a field assistant in the Kimberley frontier region of northwest Australia gathering behavioral data on Gouldian finches (*Chloebia gouldiae*), a strikingly beautiful bird clad in purple, yellow, green, blue, and white plumage. One of the things that drew Hamilton to the finch project was that for much of the twentieth century, Gouldian finch populations had been in decline, the result of habitat loss at the hand of humans, as well as disease, and Hamilton thought it interesting to work on a project that tied behavior to both disease and conservation biology.

As his time with the finches came to a close in 2015, Hamilton's interests in behavior, disease, and conservation biology were jelling, and he applied to the PhD program at the University of Tasmania in Hobart to work with Rodrigo Hamede. Hamede had done work on DFTD in devils that included a very early social network analysis that looked at contact between devils. Hamilton's dissertation would expand on that early research, using new social network tools and combining state-of-the-art technology and long-term behavioral data sets.

At about 60 centimeters long and weighing about 13 kilograms as adults, Tasmanian devils are the largest living marsupial carnivores and have some of the most powerful jaws of any mammal. Devils are most active at night and tend to spend their days in hollow logs. Much of the time they are a fairly solitary lot, but that changes during the mating season when males guard their mates from being usurped by rivals. And they don't do much else: "[Devils] have this crazy mating system," Hamilton says, "where they just do nothing but mate for that period. They don't really eat; they don't take care of themselves properly at all."

Guarding a mate involves a male displaying all sorts of aggressive behaviors toward rivals. Displays sometimes escalate to full-fledged fights that include dangerous biting that can, if the biting party is infected, spread DFTD, which in part explains why DFTD is much more common among males. The mating system also helps explain something else that, on the surface, seems puzzling, which is that females who get DFTD are in fact the individuals who were healthiest before getting infected. In most cases, animals in poor condition are the ones most susceptible to infection. But because the healthiest female devils are the ones most worth guarding as mates, guarded (and to that point healthy) females are those most likely to be bitten by males during courtship mate guarding, and so potentially infected, during the mating season.[7]

What Hamilton focused on was this: How did DFTD, especially the progression of DFTD tumors, affect the social networks of the devils? He knew from prior work that as tumor size grew, a devil's condition weakened, and its ability to compete for food and other resources was compromised. How did all of this affect social network dynamics? To answer those questions, he needed to look at a population that had a sufficient number of both healthy and sick individuals to make a legitimate comparison. That population of devils, numbering just under two dozen adults, resides near Smithton in northwestern Tasmania, an eight-hour drive northwest of the University of Tasmania.

Hamilton and his colleagues rented a campground owned by a local church and began a six-month study on DFTD progression and social networks in the devils. All adult devils in the Smithton population had already been collared with proximity sensors. For the devils, these talking collars turned on when two devils were within 30 centimeters of each other—close enough that fighting and biting were a real possibility. To get access to the data being logged on the collars, Hamilton needed to periodically capture the devils and download the data from those collars to build a social network based on proximity. But capturing devils was about more than just data downloads: it allowed Hamilton to assess the progression of tumors in infected individuals.

For about two weeks per month, each night Hamilton placed out forty traps—PVC pipes with a door that closed when the devil went in to get the yummy sheep meat bait. The goal was to capture each adult monthly, and Hamilton was quite successful. "Devils have a reputation for being these fearsome animals, going around ravaging the Tasmanian bush," Hamilton says with a smile. "They're actually very, very timid, kind of shy animals." But those jaws are dangerous enough that he was extra careful getting a devil out of the trap and into a sack, where—as long as the devil was blindfolded, and Hamilton made *very* sure it was—it went temporarily limp. He then checked the animal for bite marks and examined its face, head, and inside its mouth for the presence of DFTD tumors, and measured the size of any tumors present.

Hamilton's study lasted for six months. At the start of the study, three devils were known to be infected with DFTD, another seven developed DFTD over the course of the six months, and twelve devils were DFTD-free the entire six months. Hamilton's capture-and-release system allowed for tracking the progression of DFTD tumors, and using the downloaded data from the talking collars, he constructed social networks based on proximity across both the mating season and the non-mating season.

The talking collars turned on 8,504 times, and Hamilton found that the network density—the ratio of observed links between devils to the number of possible links—was much greater in the mating season, when males are tenaciously guarding females. Hamilton also found that DFTD progression in infected devils affected social network dynamics. He had divided tumor size into four categories, going from smaller to larger, and for each move from one category up to the next, devils were about 15% less likely to form a link with others in their network.

Aside from a short stretch of time that we will return to in a moment, DFTD-infected devils were far less interactive in their network than non-infected individuals. Devils with DFTD not only had fewer connections but were also less likely to be a hub in their network than animals that were DFTD-free. There are at least two ways this might happen: infected devils may be seeking less interaction with others, or others in the network may be actively avoiding infected devils. Ostracism is, at least in principle, possible since DFTD tumors on the face and head of infected individuals are obvious, and DFTD infection likely changes the olfactory cues a devil produces, making it even more obvious who is infected and who is not. Two lines of evidence, however, suggest that devils are not actively avoiding infected network members. For one thing, long-term pairwise interactions don't change when one member contracts DFTD and becomes symptomatic. For another, if ostracism was occurring, then Hamilton should have detected cliques, some made up of infected individuals and others composed of uninfected individuals, but no such cliques were found when he searched for them in his network analysis. Which is all to say that, for the most part, devils with DFTD tend to avoid others in their network, both in the non-mating season and most of the mating season. But not the entire mating season.

As the mating season progressed, the difference in the number of associates and in-betweenness (how many of the shortest paths link pairs of individuals in a network) in DFTD and DFTD-free individuals decreased. At first, Hamilton was puzzled by this. Why would DFTD-infected individuals become more embedded in networks during this one period? Then it struck him that because devils are what he refers to as energetically "all in" during the mating season, healthier, DFTD-free individuals, who would be most likely early on to do the mating and mate guarding, might be getting worn down by the latter part of the mating season. This then opened the door, at least a little, for DFTD-infected individuals to obtain mating opportunities, and hence to be more deeply embedded in the network.[8]

Eventually DFTD kills infected devils, but in the process, it wreaks havoc with devil social life, spreading a transmittable cancer along devil social networks and changing network dynamics as it does so.

On the ground, in the air, and in the water, animals are embedded in social networks. From devils in a remote corner of Tasmania and lemurs on Madagascar to yellow-bellied marmots in the Rocky Mountains and house mice in Switzerland, social networks impact almost every aspect of animal sociality.

It's a wondrous, complex networked world out there.

Afterword

It's time to scratch off another item from the "what makes humans unique" list. The more we learn about social networks in nonhumans, the more we realize that those networks are remarkably complex. Everywhere, and in every context, animals are embedded in social networks.

Like all scientists, I take out my crystal ball to predict the future with much trepidation. Still, certain trends peek through when I wipe back the fog. No area in the field of animal behavior is more integrally tied to technology than the study of social networks. Passive integrated transponder (PIT) tags, radio-frequency identification (RFID) antennae, GPS tracking, proximity sensors, and even satellites have already revolutionized the study of networks by providing the sort of data that just a few years back would have been the stuff of dreams. As data-collection technology continues to move forward at a breakneck pace, I have no doubt that my colleagues studying nonhuman social networks will constantly be adopting and adapting new technologies to provide an even deeper understanding of those networks. Technology will never replace the good old-fashioned on-the-ground fieldwork that we saw so much of throughout the course of this book, but it can complement and advance it.

New hardware is important, but it's only a tool, and without new conceptual and theoretical breakthroughs, that tool tells us only so much. Fortunately, animal behaviorists are constantly coming up with new ways to think about nonhuman social networks. As just one example, the work on multilayer social network analysis that we touched on in the discussion of badgers is only in its infancy. In time, that model and its descendants will mature, and as that happens, the ability to track animal social networks across time, across species, and across different behaviors will mature as well. But it's more than just tracking in new ways: multilayer social network analysis broadens how we conceptualize the interaction of the different networks in which animals are embedded. The fact that these models are largely being built by a new generation of mathematically savvy animal behaviorists, not only adept at building models from scratch, but also at importing and modifying models first developed to study human networks, only makes it more exciting.

The recognition that social networks are important in animal sociality has facilitated new lines of collaborative research. Historically, animal behavior, disease ecology, and conservation biology have been distinct, and often disjointed, research enterprises. The study of social network analysis is an ideal bridge between them. We've seen hints of this in studies of networks in endangered Barbary macaques, northern muriquis, and Verreaux's sifakas, as well as in animal networks that serve as microbial highways in Tasmanian devils, badgers, and savannah baboons, and in time our understanding of the connections between behavior, disease, and conservation will only increase. And all three disciplines will be the better for it.

Studies of animal behavior, both from a cost-benefit as well as a cognitive perspective, long predates explicit work on animal social networks. Now that we know that in many species, and many contexts, everything from foraging, mating, power, safety, travel/migration, communication, and cultural transmission take place in networked societies, it raises the question of whether we need to develop new ways to think about how to measure (direct and indirect) costs and benefits, and what sort of cognitive machinery might be in place to deal with life in a networked world. Time will tell.

Acknowledgments

My own social network contains many cliques. Each has been important in putting this book together, but none more so than the clique of colleagues I interviewed, at length, for the studies in this book. I'm deeply indebted to the following generous folks for interviews (via Zoom, Skype, WhatsApp, phone, and email) about their work on social networks in nonhumans: Jenny Allen, Lucy Aplin, Elizabeth Archie, Emily Best, Daniel Blumstein, Lauren Brent, Jakob Bro-Jørgensen, Charlotte Canteloup, Patrick Chiyo, Darren Croft, Ke Deng, Cody Dey, Robin Dunbar, Julian Evans, David Fisher, Andrea Flack, Vince Formica, Gabriella Gall, Benjamin Geffory, Anne Goldizen, David Hamilton, Matt Hasenjager, Cat Hobaiter, Amiyaal Ilany, Barbara König, Julia Lehmann, David Lusseau, Janet Mann, Nkabeng Maruping-Mzileni, David McDonald, Richard McFarland, Julie Morand-Ferron, Kevin Oh, Yannis Papastamatiou, Amanda Perofsky, Rob Perryman, Dominique Potvin, Simon Ripperger, Anna Roberts, Sam Roberts, Miho Saito, Masaki Shimada, Andy Sih, Matthew Silk, Paulo Simões-Lopes, Cedrick Sueur, Camille Testard, Marcos Tokuda, Alexander Wilson, and Dong-Po Xia. They made me feel as if I was there with them in the Dampier Strait, at Amboseli, on the mountains of Rùm, flying with storks, and on and on. I hope my channeling of their tales made you feel the same way.

My editorial clique at the University of Chicago Press is

brimming with talent. My editor Joe Calamia is as good as they come, and his input at every stage in the process was invaluable. It only helps matters that he is such a nice fellow. The family and friend cliques in my social network were, as ever, not only patient with me as I took time away from them researching and writing, but insightful in their comments on the manuscript. Special thanks are due to Dana Dugatkin, Aaron Dugatkin, Henry Bloom, Michael Sims, Lena P., and Red C. for all their help along the way.

Notes

Preface

1. L. A. Dugatkin, "Dynamics of the TIT FOR TAT Strategy during Predator Inspection in Guppies," *Behavioral Ecology and Sociobiology* 29 (1991): 127–32; L. A. Dugatkin, "Sexual Selection and Imitation: Females Copy the Mate Choice of Others," *American Naturalist* 139 (1992): 1384–89.

2. R. L. Earley and L. A. Dugatkin, "Eavesdropping on Visual Cues in Green Swordtail (*Xiphophorus helleri*) Fights: A Case for Networking," *Proceedings of the Royal Society of London* 269 (2002): 943–52.

Chapter One

1. J. Eckermann, *Conversations of Goethe* (Boston: Da Capo Press, 1998), 117.

2. M. J. Kessler and R. G. Rawlins, "A 75-Year Pictorial History of the Cayo Santiago Rhesus Monkey Colony," *American Journal of Primatology* 78 (2016): 6–43.

3. D. Croft, R. James, and J. Krause, *Exploring Animal Social Networks* (Princeton, NJ: Princeton University Press, 2008).

4. L. J. Brent, A. MacLarnon, M. L. Platt, and S. Semple, "Seasonal Changes in the Structure of Rhesus Macaque Social Networks," *Behavioral Ecology and Sociobiology* 67 (2013): 349–59.

5. S. N. Ellis, N. Snyder-Mackler, A. Ruiz-Lambides, M. L. Platt, and L. J. Brent, "Deconstructing Sociality: The Types of Social Connections That Predict Longevity in a Group-Living Primate," *Proceedings of the Royal Society of London* 286 (2019), https://doi.org/10.1098/rspb.2019.1991. The tendency to interact with friends who have lots of friends may have a genetic underpinning, and though it is not completely clear what genes are involved, there is some evidence the genes associated with the production of serotonin may play a role: L. J. Brent, S. R. Heilbronner, J. E. Horvath, J. Gonzalez-Martinez, A. Ruiz-Lambides, A. G. Robinson, H. P. Skene, and M. L. Platt, "Genetic Origins of Social Networks in Rhesus Macaques," *Scientific Reports* 3 (2013), https://doi.org/10.1038/srep01042; L. J. Brent, "Friends of Friends: Are Indirect Connections in Social Networks Important to Animal Behaviour?" *Animal Behaviour* 103 (2015): 211–22.

6. Sometimes a pair would also sing a *wit* call, but *toledos* were more common.

7. D. B. McDonald, "A Spatial Dance to the Music of Time in the Leks of Long-Tailed Manakins," *Advances in the Study of Behavior* 42 (2010): 55–81; D. B. McDonald and W. K. Potts, "Cooperative Display and Relatedness among Males in a Lek-Mating Bird," *Science* 266 (1994): 1030–32; J. M. Trainer and D. B. McDonald, "Vocal Repertoire of the Long-Tailed Manakin and Its Relation to Male-Male Cooperation," *Condor* 95 (1993): 769–81; J. M. Trainer and D. B. McDonald, "Singing Performance, Frequency Matching and Courtship Success of Long-Tailed Manakins (*Chiroxiphia linearis*)," *Behavioral Ecology and Sociobiology* 37 (1995): 249–54.

8. D. B. McDonald, "Predicting Fate from Early Connectivity in a Social Network," *Proceedings of the National Academy of Sciences* 104 (2007): 10910–14; D. B. McDonald, "Young-Boy Networks without Kin Clusters in a Lek-Mating Manakin," *Behavioral Ecology and Sociobiology* 63 (2009): 1029–34; A. J. Edelman and D. B. McDonald, "Structure of Male Cooperation Networks at Long-Tailed Manakin Leks," *Animal Behaviour* 97 (2014): 125–33. Increased success also included an increase in the number of dances for a female or the number of copulations (McDonald, "Young-Boy Networks," 2007).

9. P. Oxford, R. Bish, K. Swing, and A. Di Fiore, *Yasuni Tiputini and the Web of Life* (Quito: Ingwe Press, 2012);.

10. M. Heindl, "Social Organization on Leks of the Wire-Tailed Manakin in Southern Venezuela," *Condor* 104 (2002): 772–79; P. Schwartz and D. W. Snow, "Display and Related Behavior of the Wire-Tailed Manakin," *Living Bird* 17 (1978): 51–78.

11. R. Dakin and T. B. Ryder, "Dynamic Network Partnerships and Social Contagion Drive Cooperation," *Proceedings of the Royal Society of London* 285 (2018): https://doi.org/10.1098/rspb.2018.1973.

12. R. Dakin and T. B. Ryder, "Reciprocity and Behavioral Heterogeneity Govern the Stability of Social Networks," *Proceedings of the National Academy of Sciences* 117 (2020): 2993–99.

13. C. Testard, M. Larson, M. M. Watowich, C. H. Kaplinsky, A. Bernau, M. Faulder, H. H. Marshall et al., "Rhesus Macaques Build New Social Connections after a Natural Disaster," *Current Biology* 31 (2021): 2299–309.

Chapter Two

1. A. Espinas, "Des sociétés animales" (PhD diss., Sorbonne, 1877), 211–13; E. D'Hombres and S. Mehdaoui, "On What Condition Is the Equation Organism-Society Valid? Cell Theory and Organicist Sociology in the Works of Alfred Espinas (1870s–80s)," *History of the Human Sciences* 25 (2012): 32–35; M. Hasenjager and L. A. Dugatkin, "Social Network Analysis in Behavioral Ecology," *Advances in the Study of Behavior* 47 (2015): 39–114; K. Wils and A. Rasmussen, "Sociology in a Transnational Perspective: Brussels, 1890–1925," *Revue Belge de Philologie et d'Histoire* 90 (2012): 1273–96.

2. J. H. Crook, "Introduction: Social Behaviour and Ethology," in *Social Behaviour in Birds and Mammals*, ed. J. H. Crook (London: Academic Press, 1970), xxi–xl; N. Tinbergen, *Social Behaviour in Animals* (London: Butler and Tanner, 1953); K. Lorenz, "The Companion in the Bird's World," *Auk* 54 (1937): 245–73; D. Lack, *Ecological Adaptations for Breeding in Birds* (London: Chapman and Hall, 1968); G. McBride, "A General Theory of Social Organization and Behaviour," University of Queensland Papers, Faculty of Veterinary Science, vol. 1 (1964): 75–110.

3. W.R. Thompson, "Social Behaviour," in *Behavior and Evolution*, ed. A. Roe and G. G. Simpson (New Haven, CT: Yale University Press, 1958), 291–310; E. O. Wilson, *Sociobiology: The New Synthesis* (Cambridge, MA: Harvard University Press, 1975).

4. D. S. Sade, "Some Aspects of Parent-Offspring and Sibling Relations in a Group of Rhesus Monkeys, with a Discussion of Grooming," *American Journal of Physical Anthropology* 23 (1965): 1–17.

5. K. Lewin, *Field Theory in Social Science* (New York: Harper, 1951); J. L. Moreno, *Who Shall Survive? A New Approach to the Problem of Human Interrelations* (Washington, DC: Nervous and Mental Disease Publishing, 1934); R. A. Hinde, "Interactions, Relationships and Social Structure," *Man* 11 (1976): 1–17; L. J. Brent, J. Lehmann, and G. Ramos-Fernandez, "Social Network Analysis in the Study of Nonhuman Primates: A Historical Perspective," *American Journal of Primatology* 73 (2011): 720–30.

6. For a few excellent, if somewhat technical, reviews of social network analysis, see J. Krause, R. James, D. Franks, and D. Croft, eds., *Animal Social Networks* (Oxford: Oxford University Press, 2015); M. Hasenjager, M. Silk, and D. N. Fisher, "Multilayer Network Analysis: New Opportunities and Challenges for Studying Animal Social Systems," *Current Zoology* 67 (2021): 45–48; Q. M. R. Webber and E. Vander Wal, "Trends and Perspectives on the Use of Animal Social Network Analysis in Behavioural Ecology: A Bibliometric Approach," *Animal Behaviour* 149 (2019): 77–87; D. P. Croft, S. K. Darden, and T. W. Wey, "Current Directions in Animal Social Networks," *Current Opinion in Behavioral Sciences* 12 (2016): 52–58; J. A. Firth, "Considering Complexity: Animal Social Networks and Behavioural Contagions," *Trends in Ecology & Evolution* 35 (2020): 100–104; D. Shizuka and A. E. Johnson, "How Demographic Processes Shape Animal Social Networks," *Behavioral Ecology* 31 (2020): 1–11; D. R. Farine, "A Guide to Null Models for Animal Social Network Analysis," *Methods in Ecology and Evolution* 8 (2017): 1309–20; S. Sosa, D. M. Jacoby, M. Lihoreau, and C. Sueur, "Animal Social Networks: Towards an Integrative Framework Embedding Social Interactions, Space and Time," *Methods in Ecology and Evolution* 12 (2021): 4–9; D. R. Farine and H. Whitehead, "Constructing, Conducting and Interpreting Animal Social Network Analysis," *Journal of Animal Ecology* 84 (2015): 1144–63.

7. A. Davis, R. E. Major, C. E. Taylor, and J. M. Martin, "Novel Tracking and Reporting Methods for Studying Large Birds in Urban Landscapes," *Wildlife Biology* 4 (2017), https://doi.org/10.2981/wlb.00307.

8. M. E. J. Newman, "The Structure and Function of Complex Networks," *SIAM Review* 45 (2003): 167–256; H. Whitehead, *Analyzing Animal Societies: Quantitative Methods for Vertebrate Social Analysis* (Chicago: University of Chicago Press, 2008); L. M. Aplin, R. E. Major, A. Davis, and J. M. Martin, "A Citizen Science Approach Reveals Long-Term Social Network Structure in an Urban Parrot, *Cacatua galerita*," *Journal of Animal Ecology* 90 (2020), https://doi.org/10.1111/1365-2656.13295.

9. A. Kershenbaum, A. Ilany, L. Blaustein, and E. Geffen, "Syntactic Structure and Geographical Dialects in the Songs of Male Rock Hyraxes," *Proceedings of the Royal Society of London* 279 (2012): 2974–81; L. Koren and E. Geffen, "Complex Calls in Male Rock Hyrax (*Procavia capensis*): A Multi-Information Distributing Channel," *Behavioral Ecology and Sociobiology* 63 (2009): 581–90; A. Ilany, A. Barocas, M. Kam, T. Ilany, and E. Geffen, "The Energy Cost of Singing in Wild Rock Hyrax Males: Evidence for an Index Signal," *Animal Behaviour* 85 (2013): 995–1001; E. R. Bar Ziv, A. Ilany, V. Demartsev, A. Barocas, E. Geffen, and L. Koren, "Individual, Social, and Sexual Niche Traits Affect Copulation Success in a Polygynandrous Mating System," *Behavioral Ecology and Sociobiology* 70 (2016): 901–12.

10. A. Barocas, A. Ilany, L. Koren, M. Kam, and E. Geffen, "Variance in Centrality within Rock Hyrax Social Networks Predicts Adult Longevity," *PLoS One* 6 (2011), https://doi.org/10.1371/journal.pone.0022375.

11. J. C. Wiszniewski, C. Brown, and L. M. Moller, "Complex Patterns of Male Alliance Formation in a Dolphin Social Network," *Journal of Mammalogy* 93 (2012): 239–50; J. Mourier, C. Brown, and S. Planes, "Learning and Robustness to Catch-and-Release Fishing in a Shark Social Network," *Biology Letters* 13 (2017), https://doi.org/10.1098/rsbl.2016.0824.

12. R. J. Perryman, S. K. Venables, R. F. Tapilatu, A. D. Marshall, C. Brown, and D. W. Franks, "Social Preferences and Network Structure in a Population of Reef Manta Rays," *Behavioral Ecology and Sociobiology* 73 (2019), https://doi.org/10.1007/s00265-019-2720-x; R. J. Perryman, "Social Organisation, Social Behaviour and Collective Movements in Reef Manta Rays" (PhD diss., Macquarie University, 2020); R. J. Perryman, M. Carpenter, E. Lie, G. Sofronov, A. D. Marshall, and C. Brown, "Reef Manta Ray Cephalic Lobe Movements Are Modulated during Social Interactions," *Behavioral Ecology and Sociobiology* 75 (2021), https://doi.org/10.1007/s00265-021-02973-x.

13. More specifically, they looked at eigenvector centrality and weighted eigenvector centrality.

14. C. Canteloup, I. Puga-Gonzalez, C. Sueur, and E. van de Waal, "The Consistency of Individual Centrality across Time and Networks in Wild Vervet Monkeys," *American Journal of Primatology* 83 (2020), https://doi.org/10.1002/ajp.23232; C. Canteloup, W. Hoppitt, and E. van de Waal, "Wild Primates Copy

Higher-Ranked Individuals in a Social Transmission Experiment," *Nature Communications* 11 (2020), https://doi.org/10.1038/s41467-019-14209-8; C. Canteloup, M. Cera, B. Barrett, and E. van de Waal, "Processing of Novel Food Reveals Payoff and Rank-Biased Social Learning in a Wild Primate," *Scientific Reports* 11 (2021), https://doi.org/10.1038/s41598-021-88857-6.

Chapter Three

1. M. Newman, D. Watts, and S. Strogatz, "Random Graph Models of Social Networks," *Proceedings of the National Academy of Sciences* 99 (2002): 2566–72; D. Lusseau, "The Emergent Properties of a Dolphin Social Network," *Proceedings of the Royal Society of London B* 270 (2003): S186–S188.

2. "Complex networks that contain": Lusseau, "The Emergent Properties of a Dolphin Social Network," S186.

3. S. Milgram, "The Small-World Problem," *Psychology Today* 2 (1967): 60–67.

4. Lusseau, "The Emergent Properties of a Dolphin Social Network."

5. D. Lusseau and M. E. J. Newman, "Identifying the Role That Animals Play in Their Social Networks," *Proceedings of the Royal Society of London* 271 (2004): S477–81; D. Lusseau, "Evidence for Social Role in a Dolphin Social Network," *Evolutionary Ecology* 21 (2007): 357–66; D. Lusseau and L. Conradt, "The Emergence of Unshared Consensus Decisions in Bottlenose Dolphins," *Behavioral Ecology and Sociobiology* 63 (2009): 1067–77; D. Lusseau, "The Short-Term Behavioral Reactions of Bottlenose Dolphins to Interactions with Boats in Doubtful Sound, New Zealand," *Marine Mammal Science* 22 (2006): 801–18; D. Lusseau, "Why Do Dolphins Jump? Interpreting the Behavioural Repertoire of Bottlenose Dolphins (*Tursiops sp.*) in Doubtful Sound, New Zealand," *Behavioural Processes* 73 (2006): 257–65.

6. R. Trivers, "The Evolution of Reciprocal Altruism," *Quarterly Review of Biology* 46 (1971): 189–226; G. Wilkinson, "Reciprocal Food Sharing in the Vampire Bat," *Nature* 308 (1984): 181–84; G. Carter, D. R. Farine, and G. S. Wilkinson, "Social Bet-Hedging in Vampire Bats," *Biology Letters* 13 (2017), https://doi.org/10.1098/rsbl.2017.0112; G. Wilkinson, "Food Sharing in Vampire Bats," *Scientific American* (February 1990): 76–82.

7. S. Ripperger, G. Carter, N. Duda, A. Koelpin, B. Cassens, R. Kapitza, D. Josic, J. Berrio-Martinez, R. A. Page, and F. Mayer, "Vampire Bats That Cooperate in the Lab Maintain Their Social Networks in the Wild," *Current Biology* 29 (2019): 4139–44.

8. S. Ripperger and G. Carter, "Social Foraging in Vampire Bats Is Predicted by Long-Term Cooperative Relationships," *PLoS Biology* 19 (2021), https://doi.org/10.1371/journal. pbio.3001366.

9. For more on the Amboseli Elephant Research Project, see https://www.elephanttrust.org/index.php/meet-the-team/item/harvey.

10. P. I. Chiyo, C. J. Moss, E. A. Archie, J. A. Hollister-Smith, and S. C. Alberts, "Using Molecular and Observational Techniques to Estimate the

Number and Raiding Patterns of Crop-Raiding Elephants," *Journal of Applied Ecology* 48 (2011): 788–96; E. Archie and P. I. Chiyo, "Elephant Behaviour and Conservation: Social Relationships, the Effects of Poaching, and Genetic Tools for Management," *Molecular Ecology* 21 (2012): 765–78.

11. P. I. Chiyo, C. J Moss, and S. C. Alberts, "The Influence of Life History Milestones and Association Networks on Crop-Raiding Behavior in Male African Elephants," *PLoS One* 7 (2012), https://doi.org/10.1371/journal.pone.0031382; P. I. Chiyo, E. A. Archie, J. A. Hollister-Smith, P. C. Lee, J. H. Poole, C. J. Moss, and S. C. Alberts, "Association Patterns of African Elephants in All-Male Groups: The Role of Age and Genetic Relatedness," *Animal Behaviour* 81 (2011): 1093–99; P. I. Chiyo, J. W. Wilson, E. A. Archie, P. C. Lee, C. J. Moss, and S. C. Alberts, "The Influence of Forage, Protected Areas, and Mating Prospects on Grouping Patterns of Male Elephants," *Behavioral Ecology* 25 (2014): 1494–504.

12. L. A. Aplin, B. Sheldon, and J. Morand-Ferron, "Milk-Bottles Revisited: Social Learning and Individual Variation in the Blue Tit (*Cyanistes caeruleus*)," *Animal Behaviour* 85 (2013): 1225–32; L. Aplin, D. R. Farine, J. Morand-Ferron, and B. Sheldon, "Social Networks Predict Patch Discovery in a Wild Population of Songbirds," *Proceedings of the Royal Society of London* 279 (2012): 4199–205.

13. S. Smith, *The Black-Capped Chickadee* (Ithaca, NY: Cornell University Press, 1992).

14. L. A. Giraldeau and T. Caraco, *Social Foraging Theory* (Princeton, NJ: Princeton University Press, 2000); M. Webster, N. Atton, W. Hoppitt, and K. N. Laland, "Environmental Complexity Influences Association Network Structure and Network-Based Diffusion of Foraging Information in Fish Shoals," *American Naturalist* 181 (2013): 235–44; M. Rafacz and J. J. Templeton, "Environmental Unpredictability and the Value of Social Information for Foraging Starlings," *Ethology* 109 (2003): 951–60.

15. T. B. Jones, L. Aplin, I. Devost, and J. Morand-Ferron, "Individual and Ecological Determinants of Social Information Transmission in the Wild," *Animal Behaviour* 129 (2017): 93–101.

16. D. S. da Rosa, N. Hanazaki, M. Cantor, P. C. Simões-Lopes, and F. G. Daura-Jorge, "The Ability of Artisanal Fishers to Recognize the Dolphins They Cooperate With," *Journal of Ethnobiology and Ethnomedicine* 16 (2006), https://doi.org/10.1186/s13002-020-00383-3; F. G. Daura-Jorge, S. N. Ingram, and P. C. Simões-Lopes, "Seasonal Abundance and Adult Survival of Bottlenose Dolphins (*Tursiops truncatus*) in a Community That Cooperatively Forages with Fishermen in Southern Brazil," *Marine Mammal Science* 29 (2013): 293–311.

17. A. M. Machado, F. G. Daura-Jorge, D. F. Herbst, P. C. Simões-Lopes, S. N. Ingram, P. V. de Castilho, and N. Peroni, "Artisanal Fishers' Perceptions of the Ecosystem Services Derived from a Dolphin-Human Cooperative Fishing Interaction in Southern Brazil," *Ocean & Coastal Management* 173 (2019): 148–56.

18. Lusseau was never involved in the data collection but was a contributing author on a number of papers.

19. F. G. Daura-Jorge, M. Cantor, S. Ingram, D. Lusseau, and P. C. Simões-Lopes, "The Structure of a Bottlenose Dolphin Society Is Coupled to a Unique Foraging Cooperation with Artisanal Fishermen," *Biology Letters* 8 (2012): 702–5; M. Cantor, L. L. Wedekin, P. R. Guimaraes, F. G. Daura-Jorge, M. R. Rossi-Santos, and P. C. Simões-Lopes, "Disentangling Social Networks from Spatiotemporal Dynamics: The Temporal Structure of a Dolphin Society," *Animal Behaviour* 84 (2012): 641–51; P. C. Simões-Lopes, F .G. Daura-Jorge, and M. Cantor, "Clues of Cultural Transmission in Cooperative Foraging between Artisanal Fishermen and Bottlenose Dolphins, *Tursiops truncatus* (Cetacea: Delphinidae)," *Zoologia* 33 (2016), https://doi.org/10.1590/S1984-4689zool-20160107; C. Bezamat, P. C. Simões-Lopes, P. V. Castilho, and F. G. Daura-Jorge, "The Influence of Cooperative Foraging with Fishermen on the Dynamics of a Bottlenose Dolphin Population," *Marine Mammal Science* 35 (2019): 825–42; B. Romeu, M. Cantor, C. Bezamat, P. C. Simões-Lopes, and F. G. Daura-Jorge, "Bottlenose Dolphins That Forage with Artisanal Fishermen Whistle Differently," *Ethology* 123 (2017): 906–15; L. A. Dugatkin and M. Hasenjager, "The Networked Animal," *Scientific American* 312 (2015): 51–55.

Chapter Four

1. A. W. Goldizen, A. R. Goldizen, D. A. Putland, D. M. Lambert, C. D. Millar, and J. C. Buchan, "'Wife-Sharing' in the Tasmanian Native Hen (*Gallinula mortierii*): Is It Caused by a Male-Biased Sex Ratio?" *Auk* 115 (1998): 528–32; A. W. Goldizen, D. A. Putland, and A. R. Goldizen, "Variable Mating Patterns in Tasmanian Native Hens (*Gallinula mortierii*): Correlates of Reproductive Success," *Journal of Animal Ecology* 67 (1998): 307–17; A. W. Goldizen, J. C. Buchan, D. A. Putland, A. R. Goldizen, and E. A. Krebs, "Patterns of Mate-Sharing in a Population of Tasmanian Native Hens, *Gallinula mortierii*," *Ibis* 142 (2000): 40–47.

2. A. J. Carter, S. L. Macdonald, V. A. Thomson, and A. W. Goldizen, "Structured Association Patterns and Their Energetic Benefits in Female Eastern Grey Kangaroos, *Macropus giganteus*," *Animal Behaviour* 77 (2009): 839–46; A. J. Carter, O. Pays, and A. W. Goldizen, "Individual Variation in the Relationship between Vigilance and Group Size in Eastern Grey Kangaroos," *Behavioral Ecology and Sociobiology* 64 (2009): 237–45.

3. T. Banda, M. W. Schwartz, and T. Caro, "Woody Vegetation Structure and Composition along a Protection Gradient in a Miombo Ecosystem of Western Tanzania," *Forest Ecology and Management* 230 (2006): 179–85.

4. J. B. Foster and A. I. Dagg, "Notes on the Biology of the Giraffe," *African Journal of Ecology* 10 (1972): 1–16; V. A. Langman, "Cow-Calf Relationships in Giraffe (*Giraffa camelopardalis giraffa*)," *Zeitschrift für Tierpsychologie* 43 (1977): 264–86; M. Saito and G. Idani, "The Role of Nursery Group Guardian Is Not Shared Equally by Female Giraffe (*Giraffa camelopardalis tippelskirchi*),"

African Journal of Ecology 56 (2018): 1049–52; M. Saito and G. Idani, "Suckling and Allosuckling Behavior in Wild Giraffe (*Giraffa camelopardalis tippelskirchi*)," *Mammalian Biology* 93 (2018): 1–4; M. Saito and G. Idani, "Giraffe Mother-Calf Relationships in the Miombo Woodland of Katavi National Park, Tanzania," *Mammal Study* 43 (2018): 11–17.

5. M. F. Saito, F. B. Bercovitch, and G. Idani, "The Impact of Masai Giraffe Nursery Groups on the Development of Social Associations among Females and Young Individuals," *Behavioural Processes* 180 (2020), https://doi.org/10.1016/j.beproc.2020.104227.

6. J. J. Elliott and R. S. Arbib, "Origin and Status of the House Finch in the Eastern United States," *Auk* 70 (1953): 31–37; Cornell Laboratory of Ornithology data at https://birdsoftheworld.org/bow/species/houfin/cur/introduction.

7. G. Hill, "Female House Finches Prefer Colourful Males: Sexual Selection for a Condition-Dependent Trait," *Animal Behaviour* 40 (1990): 563–72; G. Hill, "Plumage Coloration Is a Sexually Selected Indicator of Male Quality," *Nature* 350 (1991): 337–39; G. Hill, *A Red Bird in a Brown Bag: The Function and Evolution of Colorful Plumage in the House Finch* (New York: Oxford University Press, 2002).

8. K. P. Oh and A. V. Badyaev, "Structure of Social Networks in a Passerine Bird: Consequences for Sexual Selection and the Evolution of Mating Strategies," *American Naturalist* 176 (2010): E80–E89.

9. E. C. Best, S. P. Blomberg, and A. W. Goldizen, "Shy Female Kangaroos Seek Safety in Numbers and Have Fewer Preferred Friendships," *Behavioral Ecology* 26 (2015): 639–46; C. S. Menz, A. W. Goldizen, S. P. Blomberg, N. J. Freeman, and E. C. Best, "Understanding Repeatability and Plasticity in Multiple Dimensions of the Sociability of Wild Female Kangaroos," *Animal Behaviour* 126 (2017): 3–16.

10. C. S. Menz, A. J. Carter, E. C. Best, N. J. Freeman, R. G. Dwyer, S. P. Blomberg, and A. W. Goldizen, "Higher Sociability Leads to Lower Reproductive Success in Female Kangaroos," *Royal Society Open Science* 7 (2020), https://doi.org/10.1098/rsos.200950. W. J. King, M. Festa-Bianchet, G. Coulson, and A. W. Goldizen, "Long-Term Consequences of Mother-Offspring Associations in Eastern Grey Kangaroos," *Behavioral Ecology and Sociobiology* 71 (2017), https://doi.org/110.1007/s00265-017-2297-1.

Chapter Five

1. Joint laying is one component of joint nesting behavior.

2. I. G. Jamieson, J. S. Quinn, P. A. Rose, and B. N. White, "Shared Paternity among Non-Relatives Is a Result of an Egalitarian Mating System in a Communally Breeding Bird, the Pukeko," *Proceedings of the Royal Society of London* 257 (1994): 271–77.

3. C. J. Dey, J. S. Quinn, A. King, J. Hiscox, and J. Dale, "A Bare-Part Ornament Is a Stronger Predictor of Dominance than Plumage Ornamentation in the Cooperatively Breeding Australian Swamphen," *Auk* 134 (2017): 317–29.

4. C. J. Dey and J. S. Quinn, "Individual Attributes and Self-Organizational Processes Affect Dominance Network Structure in Pukeko," *Behavioral Ecology* 25 (2014): 1402–8; J. L. Craig, "The Behaviour of the Pukeko, *Porphyrio porphyrio melanotus*," *New Zealand Journal of Zoology* 4 (1977): 413–33.

5. L. A. Dugatkin, *Power in the Wild: The Subtle and Not-So-Subtle Ways Animals Strive for Control over Others* (Chicago: University of Chicago Press, 2022).

6. R. A. Rodriguez-Munoz, A. Bretman, J. Slate, C. A. Walling, and T. Tregenza, "Natural and Sexual Selection in a Wild Insect Population," *Science* 328 (2010): 1269–72.

7. D. N. Fisher, R. Rodriguez-Munoz, and T. Tregenza, "Dynamic Networks of Fighting and Mating in a Wild Cricket Population," *Animal Behaviour* 155 (2019): 179–88; D. N. Fisher, R. Rodriguez-Munoz, and T. Tregenza, "Wild Cricket Social Networks Show Stability across Generations," *BMC Evolutionary Biology* 16 (2016), https://doi.org/10.1186/s12862-016-0726-9; D. N. Fisher, R. Rodriguez-Munoz, and T. Tregenza, "Comparing Pre- and Post-Copulatory Mate Competition Using Social Network Analysis in Wild Crickets," *Behavioral Ecology* 27 (2016): 912–19; A. Bretman and T. Tregenza, "Measuring Polyandry in Wild Populations: A Case Study Using Promiscuous Crickets," *Molecular Ecology* 14 (2005): 2169–79.

8. "Thirty years ago": R. Dunbar, "Dunbar's Number," The Conversation, May 12, 2021, https://theconversation.com/dunbars-number-why-my-theory-that-humans-can-only-maintain-150-friendships-has-withstood-30-years-of-scrutiny-160676; R. I. Dunbar, "Neocortex Size as a Constraint on Group Size in Primates," *Journal of Human Evolution* 20 (1992): 469–93; H. Kudo and R. I. Dunbar, "Neocortex Size and Social Network Size in Primates," *Animal Behaviour* 62 (2001): 711–22; R. I. Dunbar, "The Social Brain Hypothesis and Human Evolution," Oxford Research Encyclopedias, Psychology, March 3, 2016, https://doi.org/10.1093/acrefore/9780190236557.013.44; R. I. Dunbar, *Friends: Understanding the Power of Our Most Important Relationship* (New York: Little, Brown, 2021). Time has proven Dunbar right about just how intelligence goats are, as studies from a number of research groups have found that goats recognize other group members, distinguish between positive and negative emotional vocalizations made by other group mates, and respond to subtle behavioral changes by humans. C. Nawroth, "Socio-Cognitive Capacities of Goats and Their Impact on Human-Animal Interactions," *Small Ruminant Research* 150 (2017): 70–75; B. Pitcher, E. F. Briefer, L. Baciadonna, and A. G. McElligott, "Cross-Modal Recognition of Familiar Conspecifics in Goats," *Royal Society Open Science* 4 (2017), https://doi.org/10.1098/rsos.160346; L. Baciadonna, E. F. Briefer, L. Favaro, and A. G. McElligott, "Goats Distinguish between Positive and Negative Emotion-Linked Vocalisations," *Frontiers in Zoology* 16 (2019), https://doi.org/10.1186/s12983-019-0323-z.

9. C. R. Stanley and R. I. Dunbar, "Consistent Social Structure and Optimal Clique Size Revealed by Social Network Analysis of Feral Goats, *Capra hircus*,"

Animal Behaviour 85 (2013): 771–79. The dynamics of power that Dunbar and Stanley uncovered are not unique to the goats of Rùm. In the Great Orme area of northern Wales, just on the other side of the estuary from where Dunbar now lives in Liverpool, there is a population of pure white Persian goats that Dunbar can almost, but not quite, see from his kitchen window. When Dunbar and Stanley ran a similar network analysis on the power structure of goats in that population, results matched those found for the goats on Rùm.

10. B. Majolo, R. McFarland, C. Young, and M. Qarro, "The Effect of Climatic Factors on the Activity Budgets of Barbary Macaques (*Macaca sylvanus*)," *International Journal of Primatology* 34 (2013): 500–514.

11. N. Ménard, "Ecological Plasticity of Barbary Macaques (*Macaca sylvanus*)," *Evolutionary Anthropology* 11 (2002): 95–100; R. McFarland and B. Majolo, "Coping with the Cold: Predictors of Survival in Wild Barbary Macaques, *Macaca sylvanus*," *Biology Letters* 9 (2013), https://doi.org/10.1098/rsbl.2013.0428.

12. J. Lehmann and C. Boesch, "Sociality of the Dispersing Sex: The Nature of Social Bonds in West African Female Chimpanzees, *Pan troglodytes*," *Animal Behaviour* 77 (2009): 377–87; J. Lehmann and C. Ross, "Baboon (*Papio anubis*) Social Complexity: A Network Approach," *American Journal of Primatology* 73 (2011): 775–89; J. Lehmann and C. Ross, "Sex Differences in Baboon Social Network Position," *Folia Primatologica* 82 (2011): 333.

13. L. A. Campbell, P. J. Tkaczynski, J. Lehmann, M. Mouna, and B. Majolo, "Social Thermoregulation as a Potential Mechanism Linking Sociality and Fitness: Barbary Macaques with More Social Partners Form Larger Huddles," *Scientific Reports* 8 (2018), https://doi.org/10.1038/s41598-018-24373-4.

14. J. Lehmann, B. Majolo, and R. McFarland, "The Effects of Social Network Position on the Survival of Wild Barbary Macaques, *Macaca sylvanus*," *Behavioral Ecology* 27 (2016): 20–28.

Chapter Six

1. W. D. Hamilton, "Geometry of the Selfish Herd," *Journal Theoretical Biology* 31 (1971): 295–311; G. Williams, *Adaptation and Natural Selection* (Princeton, NJ: Princeton University Press, 1966); R. Pulliam, "On the Advantages of Flocking," *Journal Theoretical Biology* 38 (1973): 419–22; R. C. Miller, "The Significance of the Gregarious Habit," *Ecology* 3 (1922): 122–26; I. Vine, "Risk of Visual Detection and Pursuit by a Predator and the Selective Advantage of Flocking Behaviour," *Journal of Theoretical Biology* 30 (1971): 405–22.

2. T. Cucchi, J. D. Vigne, and J. C. Auffray, "First Occurrence of the House Mouse (*Mus musculus domesticus* Schwarz, 1943) in the Western Mediterranean: A Zooarchaeological Revision of Subfossil Occurrences," *Biological Journal of the Linnean Society* 84 (2005): 429–45; T. Cucchi, J. D. Vigne, J. C. Auffray, P. Croft, and E. Peltenburg, "Passive Transport of the House Mouse (*Mus musculus domesticus*) to Cyprus at the Early Preceramic Neolithic (Late 9th and 8th Millennia Cal. BC)," *Comptes Rendus Palevol* 1 (2002): 235–41.

3. B. König and A. K. Lindholm, "The Complex Social Environment of Female House Mice (*Mus domesticus*)," in *Evolution in Our Neighbourhood: The House Mouse as a Model in Evolutionary Research*, ed. M. Macholán, S. J. Baird, P. Munclinger, and J. Piálek (Cambridge: Cambridge University Press 2012), 114–34; B. König, A. K. Lindholm, P. C. Lopes, A. Dobay, S. Steinert, and F. Jens-Uwe Buschmann, "A System for Automatic Recording of Social Behavior in a Free-Living Wild House Mouse Population," *Animal Biotelemetry* 3 (2015), https://doi.org/10.1186/s40317-015-0069-0.

4. J .C. Evans, J. I. Liechti, B. Boatman and B. König, "A Natural Catastrophic Turnover Event: Individual Sociality Matters Despite Community Resilience in Wild House Mice," *Proceedings of the Royal Society of London* 287 (2020), https://doi.org/10.1098/rspb.2019.2880.

5. J. Johnson, "A Brief History of the Rocky Mountain Biological Laboratory," *Colorado Magazine* 2 (1962): 81–103.

6. *Marmota flaviventer* was formerly *Marmota flaviventris*. D. T. Blumstein, "Yellow-Bellied Marmots: Insights from an Emergent View of Sociality," *Philosophical Transactions of the Royal Society* 368 (2013), https://doi.org/10.1098/rstb.2012.0349; D. T. Blumstein, T. W. Wey, and K. Tang, "A Test of the Social Cohesion Hypothesis: Interactive Female Marmots Remain at Home," *Proceedings of the Royal Society of London* 276 (2009): 3007–12.

7. R. Sagarin, *Learning from the Octopus: How Secrets from Nature Can Help Us Fight Terrorist Attacks, Natural Disasters, and Disease* (New York: Basic Books, 2012).

8. Noisier calls have higher entropy: Blumstein, Wey, and Tang, "A Test of the Social Cohesion Hypothesis"; H. Fuong and D. T. Blumstein, "Social Security: Less Socially Connected Marmots Produce Noisier Alarm Calls," *Animal Behaviour* 154 (2019): 131–36; D. T. Blumstein, H. Fuong, and E. Palmer, "Social Security: Social Relationship Strength and Connectedness Influence How Marmots Respond to Alarm Calls," *Behavioral Ecology and Sociobiology* 7 (2017), https://doi.org/10.1007/s00265-017-2374-5. The centrality measure I discuss from Blumstein, Fuong, and Palmer is called "closeness centrality." This is slightly different than eigenvector centrality.

9. A. P. Montero, D. M. Williams, J. G. Martin, and D. T. Blumstein, "More Social Female Yellow-Bellied Marmots, *Marmota flaviventer*, Have Enhanced Summer Survival," *Animal Behaviour* 160 (2020): 113–19; Blumstein, Wey, and Tang, "A Test of the Social Cohesion Hypothesis"; D. Van Vuren and K. B. Armitage, "Survival and Dispersing of Philopatric Yellow-Bellied Marmots: What Is the Cost of Dispersal?" *Oikos* 69 (1994): 179–81. The effect of social network position on various attributes of life history is complex, and the discussion in the text touches on only one aspect of networks and life history. For more, see T. W. Wey and D. T. Blumstein, "Social Attributes and Associated Performance Measures in Marmots: Bigger Male Bullies and Weakly Affiliating Females Have Higher Annual Reproductive Success," *Behavioral Ecology and Sociobiology* 66 (2012): 1075–85; W. J. Yang, A. A. Maldonado-Chaparro, and D. T. Blumstein,

"A Cost of Being Amicable in a Hibernating Mammal," *Behavioral Ecology* 28 (2017): 11–19; D. T. Blumstein, D. M. Williams, A. N. Lim, S. Kroeger, and J. G. A. Martin, "Strong Social Relationships Are Associated with Decreased Longevity in a Facultatively Social Mammal," *Proceedings of the Royal Society of London* 85 (2018), https://doi.org/10.1098/rspb.2017.1934.

10. J. Bro-Jørgensen, D. W. Franks, and K. Meise, "Linking Behaviour to Dynamics of Populations and Communities: Application of Novel Approaches in Behavioural Ecology to Conservation," *Philosophical Transactions of the Royal Society* 374 (2019), 758–73, https://doi.org/10.1098/rstb.2019.0008; J. Szymkowiak and K. A. Schmidt, "Deterioration of Nature's Information Webs in the Anthropocene," *Oikos* (2021), https://doi.org/10.1111/oik.08504.

11. K. Stears, M. H. Schmitt, C. C. Wilmers, and A. M. Shrader, "Mixed-Species Herding Levels the Landscape of Fear," *Proceedings of the Royal Society of London* 287 (2020), https://doi.org/10.1098/rspb.2019.2555; also see M. H. Schmitt, K. Stears, and A. M. Shrader, "Zebra Reduce Predation Risk in Mixed-Species Herds by Eavesdropping on Cues from Giraffe," *Behavioral Ecology* 27 (2016): 1073–77; J. Bro-Jørgensen and W. M. Pangle, "Male Topi Antelopes Alarm Snort Deceptively to Retain Females for Mating," *American Naturalist* 176 (2010): E33–E39.

12. K. D. Meise, D. W. Franks, and J. Bro-Jørgensen, "Multiple Adaptive and Non-Adaptive Processes Determine Responsiveness to Heterospecific Alarm Calls in African Savannah Herbivores," *Proceedings of the Royal Society of London* 285 (2018), https://doi.org/10.1098/rspb.2017.2676.

13. M. Lovschal, P. K. Bocher, J. Pilgaard, I. Amoke, A. Odingo, A. Thuo, and J. C. Svenning, "Fencing Bodes a Rapid Collapse of the Unique Greater Mara Ecosystem," *Scientific Reports* 7 (2017), https://doi.org/10.1038/srep41450; R. Holdo, J. M. Fryxell, A. R. Sinclair, A. Dobson, and R. D. Holt, "Predicted Impact of Barriers to Migration on the Serengeti Wildebeest Population," *PLoS One* 6 (2011), https://doi.org/10.1371/journal.pone.0016370; K. D. Meise, W. Franks, and J. Bro-Jørgensen, "Using Social Network Analysis of Mixed-Species Groups in African Savannah Herbivores to Assess How Community Structure Responds to Environmental Change," *Philosophical Transactions of the Royal Society B* 374 (2019), https://doi.org/10.1098/rstb.2019.0009; A. Dobson, M. Borner, and T. Sinclair, "Road Will Ruin Serengeti," *Nature* 467 (2010): 272–73.

14. High tie groups—an average of thirty-eight ties for females and thirty-two ties for males. Low tie groups—an average of nine ties for females and eight ties for males; P. C. Lopes, and B. König, "Wild Mice with Different Social Network Sizes Vary in Brain Gene Expression," *BMC Genomics* 21 (2020), https://doi.org/10.3389/fnbeh.2020.00010; P. C. Lopes, H. D. Esther, M. K. Carlitz, and B. König, "Immune-Endocrine Links to Gregariousness in Wild House Mice," *Frontiers in Behavioral Neuroscience* 14 (2020), https://doi.org/10.3389/fnbeh.2020.00010. For more on the extracellular matrix, see A. Dityatev, M. Schachner, and P. Sonderegger, "The Dual Role of the Extracellular

Matrix in Synaptic Plasticity and Homeostasis," *Nature Reviews Neuroscience* 11 (2010): 735–46; O. P. Senkov, L. Andjus, L. Radenovic, E. Soriano, and A. Dityatev, "Neural ECM Molecules in Synaptic Plasticity, Learning, and Memory," in *Brain Extracellular Matrix in Health and Disease*, ed. A. Dityatev, B. Wehrle-Haller, and A. Pitkanen (Amsterdam: Elsevier Science, 2014), 53–80.

Chapter Seven

1. A. Vasiliev, "Pero Tafur: A Spanish Traveler of the Fifteenth Century and His Visit to Constantinople, Trebizond, and Italy," *Byzantion* 7 (1932): 75–122; E. D. Ross and E. Power, eds., *Pero Tafur: Travels and Adventures (1435–1439)* (New York: Harper and Brothers, 1926).

2. Today a distinction is sometimes made between homing and carrier pigeons, but that distinction did not exist in Tafur's day. In any case, what he calls a carrier pigeon is what today we would today call a homing pigeon.

3. W. Wiltschko and R. Wiltschko, "Homing Pigeons as a Model for Avian Navigation?" *Journal of Avian Biology* 48 (2017): 66–74; A. Flack, M. Akos, M. Nagy, T. Vicsek, and D. Biro, "Robustness of Flight Leadership Relations in Pigeons," *Animal Behaviour* 86 (2013): 723–32; I. B. Watts, M. Pettit, M. Nagy, T. B. de Perera, and D. Biro, "Lack of Experience-Based Stratification in Homing Pigeon Leadership Hierarchies," *Royal Society Open Science* 3 (2016), https://doi.org/10.1098/rsos.150518; B. Z. Pettit, M. Akos, T. Vicsek, and D. Biro, "Speed Determines Leadership and Leadership Determines Learning during Pigeon Flocking," *Current Biology* 25 (2015): 3132–37. To calculate at how often bird A was leading and bird B was following A (by responding to a directional change), they calculated a directional correlation delay.

4. A. Flack, M. Nagy, W. Fiedler, I. Couzin, I. D. and M. Wikelski, "From Local Collective Behavior to Global Migratory Patterns in White Storks," *Science* 360 (2018): 911–14; A. Flack, W. Fiedler, J. Blas, I. Pokrovsky, M. Kaatz, M. Mitropolsky, K. Aghababyan et al., "Costs of Migratory Decisions: A Comparison across Eight White Stork Populations," *Science Advances* 2 (2016), https://doi.org/10.1126/sciadv.1500931; A. Berdahl, A. Kao, A. Flack, P. Westley, E. Codling, I. Couzin, A. Dell, and D. Biro, "Collective Animal Navigation and Migratory Culture: From Theoretical Models to Empirical Evidence," *Philosophical Transactions of the Royal Society* 373 (2018), https://doi.org/10.1098/rstb.2017.0009.

5. "Mount Huangshan," UNESCO, whc.unesco.org/en/list/547/en.unesco.org/biosphere/aspac/huangshan.

6. J. H. Li and P. M. Kappeler, "Social and Life History Strategies of Tibetan Macaques at Mt. Huangshan," in *The Behavioral Ecology of the Tibetan Macaque*, J. H. Li, L. Sun, and P. Kappeler (Berlin: Springer Open, 2020), 17–45; C. P. Xiong and Q. S. Wang, "Seasonal Habitat Used by Tibetan Monkeys," *Acta Theriologica Sinica* 8 (1988): 176–83; J. H. Li, The Tibetan Macaque Society: A Field Study (Hefei: Anhui University Press, 1999).

7. D. P. Xia, R. C. Kyes, X. Wang, B. H. Sun, L. X. Sun, and J. H. Li, "Grooming Networks Reveal Intra- and Inter-Sexual Social Relationships in *Macaca* *thibetana*," *Primates* 60 (2019): 223–32; A. K. Rowe, J. H. Li, L. X. Sun, L. K. Sheeran, R. S. Wagner, D. P. Xia, D. A. Uhey, and R. Chen, "Collective Decision Making in Tibetan Macaques: How Followers Affect the Rules and Speed of Group Movement," *Animal Behaviour* 146 (2018): 51–61; X. Wang, L. X. Sun, L. K. Sheeran, B. H. Sun, Q. X. Zhang, D. Zhang, D. P. Xia, and J. H. Li, "Social Rank versus Affiliation: Which Is More Closely Related to Leadership of Group Movements in Tibetan Macaques (*Macaca thibetana*)?" *American Journal of Primatology* 78 (2016): 816–24; G. P. Fratellone, J. H. Li, L. K. Sheeran, R. S. Wagner, X. Wang, and L. X. Sun, "Social Connectivity among Female Tibetan Macaques (*Macaca thibetana*) Increases the Speed of Collective Movements," *Primates* 60 (2019): 183–89.

8. K. Strier, *Faces in the Forest: The Endangered Muriqui Monkeys of Brazil* (Cambridge, MA: Harvard University Press, 1999).

9. F. R. De Melo, J. P. Boubli, R. A. Mittermeier, L. Jerusalinsky, F. P. Tabacow, D. S. Ferraz, and M. Talebi, Northern Muriqui, *Brachyteles hypoxanthus*, IUCN Red List of Threatened Species, 2021, https://www.iucnredlist.org/fr/species/2994/191693399; E. Veado, "Caracterização da Feliciano Miguel Abdala," 2002, http://www.preservemuriqui.org.br/ing/artigos/caracterizacaorppn.pdf; J. P. Boubli, F. Couto-Santos, and K. B. Strier, "Structure and Floristic Composition of One of the Last Forest Fragments Containing the Northern Muriquis (*Brachyteles hypoxanthus*) Primates," *Ecotropica* 17 (2012): 53–69.

10. K. B. Strier, J. P. Boubli, C. B Possamai, and S. L. Mendes, "Population Demography of Northern Muriquis (*Brachyteles hypoxanthus*) at the Estacao Biologica de Caratinga/Reserva Particular do Patrimonio Natural-Feliciano Miguel Abdala, Minas Gerais, Brazil," *American Journal of Physical Anthropology* 130 (2006): 227–37.

11. M. Tokuda, J. P. Boubli, I. Mourthe, P. Izar, C. B. Possamai, and K. B. Strier, "Males Follow Females during Fissioning of a Group of Northern Muriquis," *American Journal of Primatology* 76 (2014): 529–38; M. Tokuda, J. P. Boubli, P. Izar, and K. B. Strier, "Social Cliques in Male Northern Muriquis *Brachyteles hypoxanthus*," *Current Zoology* 58 (2012): 342–52.

12. That data was eventually uploaded to a Movebank, an open access massive database for migration patterns of hundreds of different species. Some of the data, including a map representation of that data, can be found at https://www.movebank.org/cms/webapp?gwt_fragment=page=studies,path=study74496970.

13. G. Peron, C. H. Fleming, O. Duriez, J. Fluhr, C. Itty, S. Lambertucci, K. Safi, E. L. Shepard, and J. M. Calabrese, "The Energy Landscape Predicts Flight Height and Wind Turbine Collision Hazard in Three Species of Large Soaring Raptor," *Journal of Applied Ecology* 54 (2017): 1895–906; M. Scacco, A. Flack, O. Duriez, M. Wikelski, and K. Safi, "Static Landscape Features Predict Uplift Locations for Soaring Birds across Europe," *Royal Society Open Science* 6 (2019), https://doi.org/10.1098/rsos.181440.

Chapter Eight

1. V. Reynolds, *The Chimpanzees of the Budongo Forest: Ecology, Behaviour, and Conservation* (Oxford: Oxford University Press, Oxford, 2005).

2. A. I. Roberts, "Emerging Language: Cognition and Gestural Communication in Wild and Language Trained Chimpanzees (*Pan Troglodytes*)" (PhD diss., University of Stirling, 2010).

3. A. I. Roberts, S. J. Vick, S. G. B. Roberts, H. M. Buchanan-Smith, and K. Zuberbühler, "A Structure-Based Repertoire of Manual Gestures in Wild Chimpanzees: Statistical Analyses of a Graded Communication System," *Evolution and Human Behavior* 33 (2012): 578–89; A. I. Roberts, S. G. B. Roberts, and S. J. Vick, "The Repertoire and Intentionality of Gestural Communication in Wild Chimpanzee," *Animal Cognition* 17 (2014): 317–36; A. I. Roberts, S. J. Vick, and H. M. Buchanan-Smith, "Usage and Comprehension of Manual Gestures in Wild Chimpanzees," *Animal Behaviour* 84 (2012): 459–70.

4. K. von Frisch, *The Dance Language and Orientation of Bees* (Cambridge, MA: Harvard University Press, 1967); Noble Prize, nobelprize.org/prizes/medicine/1973/frisch/facts/; T. Seeley, *Honeybee Ecology: A Study of Adaptation in Social Life* (Princeton, NJ: Princeton University Press, 1985).

5. "a unique form of behavior": T. Seeley, *Honeybee Ecology* (Princeton, NJ: Princeton University Press, 2014), 83; T. Seeley, "Progress in Understanding How the Waggle Dance Improves the Foraging Efficiency of Honeybee Colonies," in *Honeybee Neurobiology and Behavior: A Tribute to Randolf Menzel*, ed. D. Eisenhardt, G. Galizia, and M. Giurfa (Berlin: Springer, 2012)

6. M. J. Hasenjager, W. Hoppitt, and L. A. Dugatkin, "Personality Composition Determines Social Learning Pathways within Shoaling Fish," *Proceedings of the Royal Society of London* 287 (2020), https://doi.org/10.1098/rspb.2020.1871; M. J. Hasenjager and L. A. Dugatkin, "Fear of Predation Shapes Social Network Structure and the Acquisition of Foraging Information in Guppy Shoals," *Proceedings of the Royal Society of London* 284 (2017), https://doi.org/10.1098/rspb.2017.2020; M. J. Hasenjager and L. A. Dugatkin, "Familiarity Affects Network Structure and Information Flow in Guppy (*Poecilia reticulata*) Shoals," *Behavioral Ecology* 28 (2017): 233–42.

7. The networks constructed from arrival times and interactions at the hive were "directed" in the sense that links in the network represented information flow from individuals collecting food at a full feeder toward unemployed workers. For the waggle dance network, these links were weighted by the number of waggle runs that were followed by the unemployed workers.

8. M. J. Hasenjager, W. Hoppitt, and E. Leadbeater, "Network-Based Diffusion Analysis Reveals Context-Specific Dominance of Dance Communication in Foraging Honeybees," *Nature Communications* 11 (2020), https://doi.org/10.1038/s41467-020-14410-0.

9. For example, the *FOXP2* gene in certain brain regions is associated with both song perception in birds and language acquisition in humans. Experimental work in young zebra finches has found that when the *FOXP2* gene is "knocked out"—deactivated—the ability to copy the song of adults is severely impaired: S. Haesler, C. Rochefort, B. Georgi, P. Licznerski, P. Osten, and C. Scharff, "Incomplete and Inaccurate Vocal Imitation after Knockdown of *FoxP2* in Songbird Basal Ganglia Nucleus Area X," *PLoS Biology* 5 (2007): 2885–97. For more on birdsong, see C. M. Aamodt, M. Farias-Virgens, and S. A. White, "Birdsong as a Window into Language Origins and Evolutionary Neuroscience," *Philosophical Transactions of the Royal Society* 375 (2020), https://doi.org/10.1098/rstb.2019.0060; M. D. Beecher, "Function and Mechanisms of Song Learning in Song Sparrows," *Advances in the Study of Behavior* 38 (2008): 167–225; W. A. Searcy and S. Nowicki, "Birdsong Learning, Avian Cognition and the Evolution of Language," *Animal Behaviour* 151 (2019): 217–27; R. C. Berwick, K. Okanoya, G. J. Beckers, and J. J. Bolhuis, "Songs to Syntax: The Linguistics of Birdsong," *Trends in Cognitive Sciences* 15 (2011): 113–21; E. Z. Lattenkamp, S. G. Horpel, J. Mengede, and U. Firzlaff, "A Researcher's Guide to the Comparative Assessment of Vocal Production Learning," *Philosophical Transactions of the Royal Society* 376 (2021), https://doi.org/10.1098/rstb.2020.0237.

10. D. A. Potvin, K. M. Parris, and R. A. Mulder, "Geographically Pervasive Effects of Urban Noise on Frequency and Syllable Rate of Songs and Calls in Silvereyes (*Zosterops lateralis*)," *Proceedings of the Royal Society of London* 278 (2011): 2464–69; D. A. Potvin, K. M. Parris, and R. A. Mulder, "Limited Genetic Differentiation between Acoustically Divergent Populations of Urban and Rural Silvereyes (*Zosterops lateralis*)," *Evolutionary Ecology* 27 (2013): 381–91; D. A. Potvin and K. M. Parris, "Song Convergence in Multiple Urban Populations of Silvereyes (*Zosterops lateralis*)," *Ecology and Evolution* 2 (2013): 1977–84.

11. C. Piza-Roca, K. Strickland, N. Kent, and C. H. Frere, "Presence of Kin-Biased Social Associations in a Lizard with No Parental Care: The Eastern Water Dragon (*Intellagama lesueurii*)," *Behavioral Ecology* 30 (2019): 1406–15; K. Strickland, R. Gardiner, A. J. Schultz, and C. H. Frere, "The Social Life of Eastern Water Dragons: Sex Differences, Spatial Overlap and Genetic Relatedness," *Animal Behaviour* 97 (2014): 53–61; C. Piza-Roca, K. Strickland, D. Schoeman, and C. H. Frere, "Eastern Water Dragons Modify Their Social Tactics with Respect to the Location within Their Home Range," *Animal Behaviour* 144 (2018): 27–36.

12. D. A. Potvin, K. Strickland, E. A. MacDougall-Shackleton, J. W. Slade, and C. H. Frere, "Applying Network Analysis to Birdsong Research," *Animal Behaviour* 154 (2019): 95–109; D. A. Potvin and S. M. Clegg, "The Relative Roles of Cultural Drift and Acoustic Adaptation in Shaping Syllable Repertoires of Island Bird Populations Change with Time Since Colonization," *Evolution* 69 (2015): 368–80.

13. The behavioral network was directional, meaning it looked at both how often chimpanzee 1 gestured to chimpanzee 2 and how often chimpanzee 1 was the recipient of gestures from chimpanzee 2. Results and "these gestures are less ambiguous" from A. I. Roberts and S. G. B. Roberts, "Wild Chimpanzees

Modify Modality of Gestures according to the Strength of Social Bonds and Personal Network Size," *Scientific Reports* 6 (2016), https://doi.org/10.1038/srep33864. Something similar may be going on with multimodal sexual signals and their many functions in bonobos (*Pan paniscus*): E. Genty, C. Neumann, and K. Zuberbühler, "Complex Patterns of Signalling to Convey Different Social Goals of Sex in Bonobos, *Pan paniscus*," *Scientific Reports* 5 (2015), https://doi.org/10.1038/srep16135.

Chapter Nine

1. M. Kawai, "Newly Acquired Precultural Behavior of the Natural Troop of Japanese Monkeys on Koshima Islet," *Primates* 6 (1965): 1–30; S. Kawamura, "The Process of Sub-Culture Propagation among Japanese Macaques," *Primates* (1959): 43–60.

2. G. Hunt, "Human-Like, Population-Level Specialization in the Manufacture of Pandanus Tools by New Caledonian Crows, *Corvus moneduloides*," *Proceedings of the Royal Society of London* 267 (2000): 403–13; B. C. Klump, J. E. van der Wal, J. H. St Clair, and R. Christian, "Context-Dependent 'Safekeeping' of Foraging Tools in New Caledonian Crows," *Proceedings of the Royal Society of London* 282 (2015), https://doi.org/10.1098/rspb.2015.0278; A. M. von Bayern, S. Danel, A. M. Auersperg, B. Mioduszewska, and A. Kacelnik, "Compound Tool Construction by New Caledonian Crows," *Scientific Reports* 8 (2018), https://doi.org/10.1038/s41598-018-33458-z.

3. The Shark Bay Dolphin Project: https://www.monkeymiadolphins.org.

4. R. Smolker, A. Richards, R. Connor, J. Mann, and P. Berggren, "Sponge Carrying by Dolphins (Delphinidae, Tursiops sp.): A Foraging Specialization Involving Tool Use?" *Ethology* 103 (1997): 454–65; J. B. Mann, B. L. Sargeant, J. J. Watson-Capps, Q. A. Gibson, M. R. Heithaus, R. C. Connor, and E. Patterson, "Why Do Dolphins Carry Sponges?" *PLoS One* 3 (2008), https://doi.org/10.1371/journal.pone.0003868; M. J. Krützen, J. Mann, M. R. Heithaus, R. C. Connor, L. Bejder, and W. B. Sherwin, "Cultural Transmission of Tool Use in Bottlenose Dolphins," *Proceedings of the National Academy of Sciences* 102 (2005): 8939–43; R. C. Connor, *Dolphin Politics in Shark Bay: A Journey of Discovery* (The Dolphin Alliance Project, 2018).

5. J. M. Mann, M. A. Stanton, E. M. Patterson, E. J. Bienenstock, and L. O. Singh, "Social Networks Reveal Cultural Behaviour in Tool-Using Using Dolphins," *Nature Communications* 3 (2012), https://doi.org/10.1038/ncomms1983.

6. J. M. Allen, M. Weinrich, W. Hoppitt, and L. Rendell, "Network-Based Diffusion Analysis Reveals Cultural Transmission of Lobtail Feeding in Humpback Whales," *Science* 340 (2013): 485–88; J. Allen, "A Trendy Tail: Cultural Transmission of an Innovative Lobtail Feeding Behaviour in the Gulf of Maine Humpback Whales, *Megaptera novaeangliae*" (Master's thesis, University of St. Andrews, 2011).

7. J. Fisher and R. Hinde, "The Opening of Milk Bottles by Birds," *British Birds* 42 (1949): 347–57; R. Hinde and J. Fisher, "Further Observations on the

Opening of Milk Bottles by Birds," *British Birds* 44 (1951): 393–96; L. Lefebvre, "The Opening of Milk Bottles by Birds: Evidence for Accelerating Learning Rates, but against the Wave-of-Advance Model of Cultural Transmission," *Behavioural Processes* 34 (1995): 43–53; L. M. Aplin, B. C. Sheldon, and J. Morand-Ferron, "Milk Bottles Revisited: Social Learning and Individual Variation in the Blue Tit, *Cyanistes caeruleus*," *Animal Behaviour* 85 (2013): 1225–32.

8. L. M. Aplin, D. R. Farine, J. Morand-Ferron, A. Cockburn, A. Thornton, and B. C. Sheldon, "Experimentally Induced Innovations Lead to Persistent Culture via Conformity in Wild Birds," *Nature* 518 (2015): 538–41; L. M. Aplin, B. C. Sheldon, and R. McElreath, "Conformity Does Not Perpetuate Suboptimal Traditions in a Wild Population of Songbirds," *Proceedings of the National Academy of Sciences* 114 (2017): 7830–37; S. Wild, M. Chimento, K. McMahon, D. R. Farine, B. C. Sheldon, and L M. Aplin, "Complex Foraging Behaviours in Wild Birds Emerge from Social Learning and Recombination of Components," *Philosophical Transactions of the Royal Society of London* 377 (2022): 13, https://doi.org/10.1098/rstb.2020.0307.

9. V. Reynolds, *The Chimpanzees of the Budongo Forest: Ecology, Behaviour, and Conservation* (Oxford: Oxford University Press, 2005); C. Hobaiter, T. Poisot, K. Zuberbühler, W. Hoppitt, and T. Gruber, "Social Network Analysis Shows Direct Evidence for Social Transmission of Tool Use in Wild Chimpanzees," *PLoS Biology* 12 (2014), https://doi.org/10.1371/journal.pbio.1001960.

Chapter Ten

1. S. A. Altmann and J. Altmann, *Baboon Ecology: African Field Research* (Chicago: University of Chicago Press, 1970).

2. E. A. Archie and J. Tung, "Social Behavior and the Microbiome," *Current Opinion in Behavioral Sciences* 6 (2015): 28–34; A. Sarkar, S. Harty, K. V. Johnson, A. H. Moeller, E. A. Archie, L. D. Schell, R. N. Carmody, T. H. Clutton-Brock, R. I. Dunbar, and P. W. Burnet, "Microbial Transmission in Animal Social Networks and the Social Microbiome," *Nature Ecology & Evolution* 4 (2020): 1020–35.

3. J. Tung, L. B. Barreiro, M. B. Burns, J. C. Grenier, J. Lynch, L. E. Grieneisen, J. Altmann, S. C. Alberts, R. Blekhman, and E. A. Archie, "Social Networks Predict Gut Microbiome Composition in Wild Baboons," *eLife* 4 (2015), https://doi.org/10.7554/eLife.05224.

4. A. C. Perofsky, R. J. Lewis, L. A. Abondano, A. Di Fiore, and L.A. Meyers, "Hierarchical Social Networks Shape Gut Microbial Composition in Wild Verreaux's Sifaka," *Proceedings of the Royal Society* 284 (2017), https://doi.org/10.1098/rspb.2017.2274; J. L. Round and S. K. Mazmanian, "The Gut Microbiota Shapes Intestinal Immune Responses during Health and Disease," *Nature Reviews Immunology* 9 (2009): 313–23; M. C. Abt and E. G. Pamer, "Commensal Bacteria Mediated Defenses against Pathogens," *Current Opinions in Immunology* 29 (2014): 16–22.

5. E. L. Campbell, A. W. Byrne, F. D. Menzies, K. R. McBride, C. M. Mc-

Cormick, M. Scantlebury, and N. Reid, "Interspecific Visitation of Cattle and Badgers to Fomites: A Transmission Risk for Bovine Tuberculosis?" *Ecology and Evolution* 9 (2019): 8479–89; R. C. Woodroffe, C. A. Donnelly, W. T. Johnston, F. J. Bourne, C. L. Cheeseman, R. S. Clifton-Hadley, D. R. Cox et al., "Spatial Association of *Mycobacterium bovis* Infection in Cattle and Badgers *Meles meles*," *Journal of Applied Ecology* 42 (2005): 852–62; C. A. Donnelly, R. Woodroffe, D. R. Cox, F. J. Bourne, C. L. Cheeseman, R. S. Clifton-Hadley, G. Wei et al., "Positive and Negative Effects of Widespread Badger Culling on Tuberculosis in Cattle," *Nature* 439 (2006): 843–46; D. M. Wright, N. Reid, W. I. Montgomery, A. R. Allen, R. A. Skuce, and R. R. Kao, "Herd-Level Bovine Tuberculosis Risk Factors: Assessing the Role of Low-Level Badger Population Disturbance," *Scientific Reports* 5 (2015), https://doi.org/10.1038/srep1306; K. R. Finn, M. J. Silk, M. A. Porter, and N. Pinter-Wollman, "The Use of Multilayer Network Analysis in Animal Behaviour," *Animal Behaviour* 149 (2019): 7–22; M. J. Hasenjager, M. Silk, and D. N. Fisher, "Multilayer Network Analysis: New Opportunities and Challenges for Studying Animal Social Systems," *Current Zoology* (2021): 45–58; J. L. McDonald, A. Robertson, and M. Silk, "Wildlife Disease Ecology from the Individual to the Population: Insights from a Long-Term Study of a Naturally-Infected European Badger Population," *Journal of Animal Ecology* 87 (2018): 101–12; H. Kruuk, *The Social Badger* (Oxford: Oxford University Press, 1989).

6. C. Rozins, M. Silk, D. P. Croft, R. J. Delahay, D. J. Hodgson, R. A. McDonald, N. Weber, and M. Boots, "Social Structure Contains Epidemics and Regulates Individual Roles in Disease Transmission in a Group-Living Mammal," *Ecology and Evolution* 8 (2018): 12044–55; M. Silk, N. Weber, L. C. Steward, R. J. Delahay, D. P. Croft, D. J. Hodgson, M. Boots, and R. A. McDonald, "Seasonal Variation in Daily Patterns of Social Contacts in the European Badger *Meles meles*," *Ecology and Evolution* 7 (2017): 9006–15; M. Silk, N. L. Weber, L. C. Steward, D. J. Hodgson, M. Boots, D. P. Croft, R. J. Delahay, and R. A. McDonald, "Contact Networks Structured by Sex Underpin Sex-Specific Epidemiology of Infection," *Ecology Letters* 21 (2018): 309–18; T. J. Roper, D. J. Shepherdson, and J. M. Davies, "Scent Marking with Faeces and Anal Secretion in the European Badger (*Meles meles*): Seasonal and Spatial Characteristics of Latrine Use in Relation to Territoriality," *Behaviour* 97 (1986): 94–117; M. Silk, J. A. Drewe, R. J. Delahay, N. Weber, L. C. Steward, J. Wilson-Aggarwal, M. Boots, D. J. Hodgson, D. P. Croft, and R. A. McDonald, "Quantifying Direct and Indirect Contacts for the Potential Transmission of Infection between Species Using a Multilayer Contact Network," *Behaviour* 155 (2018): 731–57.

7. J. S. Welsh, "Contagious Cancer," *Oncologist* 16 (2011): 1–4; E. A. Ostrander, B. W. Davis, and G. K. Ostrander, "Transmissible Tumors: Breaking the Cancer Paradigm," *Trends in Genetics* 32 (2016): 1–15; A. M. Dujon, R. A. Gatenby, G. Bramwell, N. MacDonald, E. Dohrmann, N. Raven, A. Schultz et al., "Transmissible Cancers in an Evolutionary Perspective," *iScience* 23 (2020), https://doi.org/10.1016/j.isci.2020.101269; R. K. Hamede, J. Bashford,

H. McCallum, and M. Jones, "Contact Networks in a Wild Tasmanian Devil (*Sarcophilus harrisii*) Population: Using Social Network Analysis to Reveal Seasonal Variability in Social Behaviour and Its Implications for Transmission of Devil Facial Tumour Disease," *Ecology Letters* 12 (2009): 1147–57; K. Wells, R. K. Hamede, D. H. Kerlin, A. Storfer, P. A. Hohenlohe, M. E. Jones, and H. I. McCallum, "Infection of the Fittest: Devil Facial Tumour Disease Has Greatest Effect on Individuals with Highest Reproductive Output," *Ecology Letters* 20 (2017): 770–78; D. G. Hamilton, M. E. Jones, E. Z. Cameron, H. McCallum, A. Storfer, P. A. Hohenlohe, and R. K. Hamede, "Rate of Intersexual Interactions Affects Injury Likelihood in Tasmanian Devil Contact Networks," *Behavioral Ecology* 30 (2019): 1087–95.

8. D. G. Hamilton, M. E. Jones, E. Z. Cameron, D. H. Kerlin, H. McCallum, A. Storfer, P. A. Hohenlohe, and R. K. Hamede, "Infectious Disease and Sickness Behaviour: Tumour Progression Affects Interaction Patterns and Social Network Structure in Wild Tasmanian Devils," *Proceedings of the Royal Society of London* 287 (2020), https://doi.org/10.6084/m9.figshare.c.5223359.

Index

Page numbers in italics and refer to figures; "gallery" indicates figures in unnumbered gallery following p. 82.

Abdala Reserve. *See* Feliciano Miguel Abdala Private Natural Heritage Reserve
ABRP. *See* Amboseli Baboon Research Project
Adams, Douglas, 123
alarm calls, 98, 100–101, 103–6
Alberts, Susan, 49–50, 164
Allen, Jenny, 146–51, 205n6
Altmann, Jeanne, 164, 206n1, 206n3
Altmann, Stuart, 164, 206n1
altruism: and cooperation, 44; reciprocal, 44, 193n6
Amboseli Baboon Research Project (ABRP), 163–68, 172
Amboseli Elephant Research Project (Kenya), 49–51, 193n9
Amboseli National Park (Kenya), 49–50, 143, 163–64
Ankoatsifaka Research Station (of University of Texas, in Kirindy Mitea National Park, Madagascar), 169, 171
antelope. *See* topi antelope (*Damaliscus lunatus*)
antennation, 128–32
Aplin, Lucy, 23–28, 53, 151–55
Archie, Elizabeth, 163–69
Artzy-Randrup, Yael, 29
Australasian swamphen. *See* pūkeko (*Porphyrio melanotus*)
Australian National Botanic Gardens, 133

Baboon Ecology: African Field Research (Altmann and Altmann), 164, 206n1
baboons, 90, 103, 143, 163–70; grooming networks of, 167–69, 172–73; savannah, 163, 184. *See also* Amboseli Baboon Research Project
badgers (*Meles meles*), 184; and cattle, 173–77; and transmission of *M. bovis*, 173–77
Badyaev, Alexander, 67–70, 196n8
Barbary Macaque Project, 87–89
Barbary macaques (*Macaca sylvanus*), 86–91, 184
bats, xii. *See also* fringe-lipped bats (*Trachops cirrhosus*); vampire bats
bees, social behavior of, 127. *See also* honeybees, cosmopolitan (*Apis mellifera*)
Bercovitch, Fred, 65, 196n5

Best, Emily, 71
betweenness: and animal behavior, 38; in mating season, 181–82; in social networks, 30, 32–34, *33*, 38; in traveling/foraging networks, 42
Big City Birds (Sydney, Australia), 24, 27–28
Biro, Dora, 110, 201nn3–4
black-capped chickadees (*Poecile atricapillus*), 43, 53–55, 58, 194n13
blue tits (*Parus caeruleus*), 152
blue wildebeests, 98, 103–4
Blumstein, Dan, 98–102
bottlenose dolphins (*Tursiops truncatus*), 39–43, 55–58, 138, 141–46, 159, 160–61
Boubli, Jean Philippe, 117, 202nn9–11
Bowdoin Scientific Station (Kent Island, Bay of Fundy), 67
Brent, Lauren, 2–5, 15–17
Brillat-Savarin, Jean Anthelme, 39
British Ornithological Society (British Trust for Ornithology [BTO]), 152
Bro-Jørgensen, Jakob, 102–6
Brown, Culum, 32, 192nn11–12
Budongo Conservation Field Station (Uganda), 124
Budongo Forest (Uganda), 123–25, 137, 157–59, 161
Budongo Forest Project, 123
Bullough, Sir George, 83

canine transmissible venereal tumor (CTVT), 177
Canteloup, Charlotte, 35–38, 192–93n14
Cantor, Mauricio, 57
Carpenter, Clarence, 2
carrier pigeons. *See* homing pigeons (*Columba livia domestica*)
Carter, Alecia, 62
Carter, Gerald, 45–49
Cayo Santiago (Puerto Rico), 1–6, 15–18, 21, 97, 189n2
centrality: and animal behavior, 38; and connectedness in networks, *37*; eigenvector and weighted eigenvector, 192n13; in grooming networks, 173; and multispecies interactions, 106; and nearest neighbor distance, 136
chickadees. *See* black-capped chickadees (*Poecile atricapillus*)
chimpanzees (*Pan troglodytes schweinfurthii*), xiii, 63, 90, 122–27, *126*, 137–38, 146, 157–61, 172, 204–5n13
Chiyo, Patrick, 49–52, 193–94nn10–11
cliques: of cockatoos, 26–27, *27*; of dolphins, 58, 146, *gallery*; of elephants, 52; of kangaroos, 72; of macaques, 114–16; of mice, 96–97; as subgroups, 20; of Tasmanian devils, 181
clustering: schematic of in social networks, *27*; in social networks, 27, 85–86; and strength, 27–28
clustering coefficients: and animal behavior, 38; and cliques, 72, 97, 145–46; defined, 27; and direct interactions, 37; and power networks, 90–91; schematic of in social networks, *27*; and social networks, 26–27, 58
Clutton-Brock, Tim, 83–84, 206n2
cockatoos, 151. *See also* sulphur-crested cockatoos
communication: and cognition, 124; and food, 132. *See also* alarm calls
communication networks, xii, 122–38, *126*, 184–85
conservation: and animal behavior, 116, 184; and anthropogenic (human-caused) factors, 106; and DFTD devastation as emergency,

177; and disease, 184; and dynamics of social networks, 106; and ethology, 50, 61
conservation biology, 40, 61, 178, 184
cooperation, and altruism, 44
cooperative behavior, ix, xii, 12, 30
Cornell Lab of Ornithology (Ithaca, NY), 67, 196n6
crickets. *See* field crickets (*Gryllus campestris*)
Croft, Darren, 2
Croze, Harvey, 50
CTVT. *See* canine transmissible venereal tumor
cultural transmission, 138, 148, 184–85; and foraging, 140–46, 149–57; and genetics, 140; and networks, 153; and tool use, 141–42, 144, 157–61, 205n4, 206n9. *See also* culture networks; information transfer
culture networks, xii, 138–61, 184–85. *See also* cultural transmission

Dakin, Roslyn, 14, 190nn11–12
Dampier Strait (Raja Ampat, West Papua), 30–34, *gallery*
Darwin, Charles, xi, 19, 41, 100
deer, red, 83
devils. *See* Tasmanian devils (*Sarcophilus harrisii*)
Dey, Cody, 76–79, 196–97nn3–4
DFTD. *See* Tasmanian devil facial tumor disease
disease: and conservation biology, 178, 184; ecology, 161, 164–65, 170, 184; infectious, 164–65, 208n8; and microbes, 165. *See also* health networks; *specific disease(s)*
dolphin-human mutualism, 56–57
dolphins, 32, *gallery*. *See also* bottlenose dolphins (*Tursiops truncatus*)
Doubtful Sound (New Zealand), 40–42, 193n5
dugongs (*Dugong dugon*), marine mammal, 142
Dunbar, Robin, 83–86, 127

Earley, Ryan, x, 189n2
eastern grey kangaroos (*Macropus giganteus*), 61–63
Ecole Nationale Forestière d'Ingénieurs (Salé, Morocco), 87
ECP. *See* extracellular protein matrix
Edward Grey Institute of Field Ornithology (EGI), Oxford University, 152
egalitarianism, in hyrax networks, 30, *gallery*
EGI. *See* Edward Grey Institute of Field Ornithology
Ein Gedi Nature Reserve (Israel), 28–29, *gallery*
Elanda Point (Australia), 62, 71–72
elephants, xii, 28, 43, 58, 63–64, 164–66; as crop raiders, 49–52
epidemiology, 163, 174
Espinas, Alfred, 19, 190n1
European white storks (*Ciconia ciconia*), 112, 119–22, 201n4
Evans, Julian, 95–97, 199n4
Exploring Animal Social Networks (Croft, James, Krause), 2, 189n3
extracellular protein matrix (ECP), 107, 200–201n14

Faces in the Forest: The Endangered Muriqui Monkeys of Brazil (Strier), 116–17, 202n8
Farine, Damien, 46, 153, 156
Feliciano Miguel Abdala Private Natural Heritage Reserve (Brazil), 112, 117, 202nn9–10
field crickets (*Gryllus campestris*), xii, 79–83, 91

finches. *See* Gouldian finches (*Chloebia gouldiae*); house finches (*Haemorhous mexicanus*); zebra finches
Fisher, David, 79–83
Fisher, James, 151–52
Flack, Andrea, 109–12, 119–22
food: and communication, 132; competing for, 179; and evolution, 58; finding, 55; and natural selection, 58; reciprocity, 44–45; sharing, 44–45, 125. *See also* lobtailed feeding and foraging
food networks, xii, 39–59, 184–85; and reciprocal altruism (cooperation), 44; and reproduction, 58–59
Francisco, Nihagnandrainy, 169–70
Franks, Daniel, 103–4
Fratellone, Gregory, 113, 115–16, 202n7
Frere, Celine, 135, 204nn11–12
fringe-lipped bats (*Trachops cirrhosus*), 45
Fuong, Holly, 100–101, 199n8

gazelles, 103–4, 164
"Geometry of the Selfish Herd" (Hamilton), 93, 198n1
giraffes, xii, 63–66, 73, 98, 103–4
goats, feral (*Capra hircus*), 83–86, 91, 197–98nn8–9
Goethe, Johann Wolfgang von, 1, 189n1
Goldizen, Anne, 61–63, 70–73
Gouldian finches (*Chloebia gouldiae*), 177–78
Great Sandy National Park (Australia), 59, 62
great tit songbirds (*Parus major*): cognitive abilities of, 53; and cultural transmission, 138, 151–57, 159, 161; foraging in Wytham Woods near Oxford, 53–55, 151–57, *156*; social networks of, 53, 138
green swordtail fish, x, 189n2
grooming networks: of baboon groups, 167–69, 172–73; centrality in, 173; and cliques, 114; of macaque monkeys, 3, 16–17, *17*, 22, 114–16, *gallery*; and microbes, 165; and microbiomes, 172–73; of vampire bats, 46–47; of vervets, 35–36
group dynamics, 19–20, 85, 90
Gruber, Thibaud, 158–59, 206n9
guppies (*Poecilia reticulata*), ix–x, 129

Hamede, Rodrigo, 178, 207–8nn7–8
Hamelin Pool Marine Nature Reserve (Australia), 142
Hamilton, David, 177–82, 208nn7–8
Hamilton, William D., 93, 198n1
Hasenjager, Matt, 128–32
health networks, 161, 163–82. *See also* disease
Henry V (Shakespeare), 93
Hill, Geoff, 68
Hinde, Robert, 21, 151–52
Hobaiter, Cat, 157–61, 206n9
homing pigeons (*Columba livia domestica*), 109–12, 119, 121, 201nn2–3
honeybees, cosmopolitan (*Apis mellifera*), 127–32, 138, *gallery*; dance behavior of, 127–31, 203n4
Hoppitt, Will, 129–30, 148–50, 159
house finches (*Haemorhous mexicanus*), 66–70, 73, *gallery*
house mice. *See* mice, house
humpback whales (*Megaptera novaeangliae*), 138, 146–51, 159, 161, 205n6
Hurricane Maria, in Cayo Santiago, 5–6, 15–18, 97
hyraxes. *See* rock hyraxes (*Procavia capensis*)

I Contain Multitudes (Yong), 163
Ifrane National Park (Morocco), 87
Ilany, Amiyaal, 28–30, 192nn9–10
impalas, 63, 98, 104
information transfer in networks, 42, 55. *See also* cultural transmission
Inkawu Vervet Project (IVP), 35
International Union for Conservation of Nature (IUCN), 87, 117, 177, 202n9
IUCN. *See* International Union for Conservation of Nature
IVP. *See* Inkawu Vervet Project
Izar, Patricia, 117, 202n11

James, Richard, 2, 189n3
Japanese macaques, 140–41, 151, 205n1
Jarman, Peter, 61–62
Jefferson, Thomas, 139
Johnson, John, 98
Journal of Pero Tafur, The, 109

kangaroos, xii, 59, 70–73, *gallery*. *See also* eastern grey kangaroos (*Macropus giganteus*)
Katavi National Park (Tanzania), 63–66, 196n4
Kawai, Masao, 140–41, 205n1
Kawamura, Shunzo, 140–41, 205n1
Kibale National Park (Uganda), 49–50
Kinloch Castle (Isle of Rùm, Scotland), 83
Kirindy Mitea National Park (Madagascar), 169–70
König, Barbara, 94–98, 106–7
Krause, Jens, 2, 189n3
Krützen, Michael, 144, 205n4

Lack, David, 152, 191n2
Leadbeater, Elli, 128–29, 203n8
Lehmann, Julia, 90
lemurs. *See* Verreaux's sifaka (*Propithecus verreauxi*)
leopards, 28, 63, 104, 113
Lewis, Rebecca, 169–71, 206n4
Li, Jin-Hua, 113, 201–2nn6–7
Life, the Universe, and Everything (Adams), 123
Life magazine, 2
lobtailed feeding and foraging, 148–51, 205n6
long-tailed manakins (*Chiroxiphia linearis*), 6–15. *See also* wire-tailed manakins (*Pipra filicauda*)
Lopes, Patricia, 106–7
Lorenz, Konrad, 127, 191n2
Lusseau, David, 39–43, 57

Maasai Amboseli Game Reserve (Kenya), 164. *See also* Masai Mara National Reserve
macaque monkeys, 1–6, 15–18, 21–22, 97, 189n2; grooming networks, 3, 16–17, *17*, 22, 114–15; life span, 3; as one of best studied of all primate species, 3; and primatology studies, 2, 87, 140; social behavior of, 113. *See also* Barbary macaques (*Macaca sylvanus*); Japanese macaques; rhesus macaque monkeys (*Macaca mulatta*); Tibetan macaques (*Macaca thibetana*)
Majolo, Bonaventura, 87, 90, 198nn10–11, 198nn13–14
Major, Richard, 24–25, 192nn7–8
manakins. *See* long-tailed manakins (*Chiroxiphia linearis*); wire-tailed manakins (*Pipra filicauda*)
manatees, 28, 142
Mandela, Nelson, 61
Mann, Janet, 141–46, 205nn4–5
manta rays, reef (*Mobula alfredi*), xii, xiii, 18, 23, 30–34, *33*, *gallery*, 192n12

Māori folklore, 75
marmots. *See* yellow-bellied marmots (*Marmota flaviventer*)
Martin, John, 24
Masai Mara National Reserve (Kenya), 98, 102–6, 133. *See also* Maasai Amboseli Game Reserve
Masai people, 49–51
Mawana Game Reserve (KwaZulu-Natal, South Africa), 23, 34–35
Maximain, Andriamampiandrisoa, 169–70
Max Planck Institute for Ornithology (Seewiesen, Germany), 112, 119
McDonald, David, 5–12
McFarland, Richard, 86–91, 198n14, 198nn10–11
McPherson, Isaac, 139
Meise, Kristine, 103–4
Mendel, Gregor, 139–40
Menz, Clementine, 71, 196nn9–10
mice: and commensal relationship with humans, 94; house (*Mus musculus*), 91, 93–98, 106–7, 182, *gallery*
microbiomes, 163–73, 206nn2–3
migration, xiii, 120; and social networks, 184–85; of white storks, 112. *See also* travel networks
Milgram, Stanley, 41–42, 193n3
monkeys. *See* macaque monkeys; northern muriqui (*Brachyteles hypoxanthus*); vervet monkeys (*Chlorocebus pygerythrus*); Yulingkeng monkeys
Montero, Anita Pilar, 101, 199n9
Monteverde Cloud Forest (Costa Rica), 7–11
Morand-Ferron, Julie, 53–55, 156
Moss, Cynthia, 50
moss sponges. *See* sponges, moss
Mount Huangshan Biosphere Reserve (China), 112–13, 201n5
Muhumuza, Geresomu, 124–25
Museum für Naturkunde (Berlin), 44
Mycobacterium bovis pathogen, 173–77

National Bison Range (Montana), 165
National Center for Ecological Analysis and Synthesis (Santa Barbara, CA), 99–100
navigation. *See* travel networks
Newman, Mark, 40
Nineteen Eighty-Four (Orwell), 75
Nobel Prize, 127
northern muriqui (*Brachyteles hypoxanthus*), 112, 116–19, 184, 202nn9–11

Oh, Kevin, 66–70
On the Origin of Species (Darwin), 41
Orwell, George, 75
Oxford Navigation Group (OxNav), 110

Palmer, Elizabeth, 101, 199n8
Perofsky, Amanda, 169–73, 206n4
Perryman, Rob, 30–34, 192n12
Petrucci, Raphaël, 19
Phillips, Daniel (Danny), 5, 15–16
Physiology of Taste, The (Brillat-Savarin), 39
pigeons. *See* homing pigeons (*Columba livia domestica*)
Platt, Michael, 2, 4, 189nn4–5
play and playing, xi–xii, 34, 36–38, 99, 113, 125
Potvin, Dominique, 132–37, 204nn10–12
power: and aggression, 77–79, 85–87, 90–91; and linear hierarchies, 79; struggles, xii, 73, 79–81, 85; and survival, 90–91
power networks, xii, 36–37, 73, 75–91, 184–85

pūkeko (*Porphyrio melanotus*), 73, 75–79, 91

Qarro, Mohamed, 87, 198n10

Raja Ampat Regency (West Papua), 23, 30
Rakatomalala, Elvis, 171
red deer, 83
reef manta rays. *See* manta rays, reef (*Mobula alfredi*)
Rendell, Luke, 148–50, 205n6
reproduction networks, xii, 59, 61–73, 184–85; and food, 58–59; and parental care, 58–59, 66, 73
Reynolds, Vernon, 123, 203n1, 206n9
rhesus macaque monkeys (*Macaca mulatta*), 1–3, 21–22; grooming networks, *17, 22, gallery*
Ripperger, Simon, 44–49, 193nn7–8
RMBL. *See* Rocky Mountain Biological Laboratory
Roberts, Anna, 123–27, 137–38, 157
Roberts, Sam, 123–27, 137–38, 157
rock hyraxes (*Procavia capensis*), 28–30, *gallery*, 192nn9–10
Rocky Mountain Biological Laboratory (Crested Butte, CO), 98–99, 199n5
Rodríguez-Muñoz, Rolando, 79–81, 197nn6–7
Rowe, Amanda, 113, 115, 202n7
Royal Botanic Garden Sydney, 23–26
Rùm National Nature Reserve (Scotland), 84
Ryder, T. Brandt, 11–14, 190nn11–12

Sade, Donald, 21–22, 191n4
safety networks, xii, 91, 93–108, 133, 184–85
Saito, Miho, 63–66, 195–96nn4–5
savannah baboons (*Papio cynocephalus*), 163, 184
savannah sparrows (*Passerculus sandwichensis*), 67
Seeley, Thomas, 127, 203nn4–5
serendipity, and social networks, xiii
Serengeti-Mara ecosystem, 103
Serengeti Reserve (Tanzania), 103
Shark Bay (Australia), 141–43, 161, 205nn3–4
Shark Bay Dolphin Project (Australia), 142–43, 205n3
Sheldon, Ben, 53, 153, 156
sifaka. *See* Verreaux's sifaka (*Propithecus verreauxi*)
silvereyes (*Zosterops lateralis*), 127, 131–37, 138, 204n10
Simões-Lopes, Paulo, 55–58
Smith, Susan, 53
Smithsonian National Zoo (Washington, DC), 164
Smithsonian Tropical Research Institute (Gamboa, Panama), 45–47, *gallery*
Smuts, Barbara, 142–43
social brain hypothesis, 83–84, 197n8
social capital, and animal behavior, 36–38
social differentiation, and animal behavior, 34, 38
sociometrics, 20–21, 23
songbirds: and genetics, 134; and human language, 132–33, 203–4n9; and social networks, 132, 137. *See also* black-capped chickadees (*Poecile atricapillus*); great tit songbirds (*Parus major*); silvereyes (*Zosterops lateralis*)
sparrows. *See* savannah sparrows (*Passerculus sandwichensis*)
sponges, 142–46; moss, 158–60
sponging behavior, 142–46, 158
Stanley, Christina, 85–86, 197–98n9
Stellwagen Bank National Marine Sanctuary (Gulf of Maine), 146–51, 161

storks. *See* European white storks (*Ciconia ciconia*)
STRI. *See* Smithsonian Tropical Research Institute
Strier, Karen, 116–17, 202nn8–11
sulphur-crested cockatoos, 18, 23–28, 27, 30, *gallery*
Sundown National Park (Australia), 62–63, 70–73, *gallery*
swamphen. *See* pūkeko (*Porphyrio melanotus*)
swordtail fish, green, x, 189n2

Tafur, Pero, 109, 201nn1–2
Tasmanian devil facial tumor disease (DFTD), 177–82
Tasmanian devils (*Sarcophilus harrisii*), xii, 177–82, 184, 207–8nn7–8
Tasmanian native hens (*Gallinula mortierii*), 61–62, 195n1
Tāwharanui Regional Park (North Island, New Zealand), 76–79
TBS. *See* Tiputini Biodiversity Station
Testard, Camille, 4, 17, 190n13
Tibetan macaques (*Macaca thibetana*), 112–16, 119, 201–2nn6–7
Tinbergen, Nikolaas, 127, 191n2
Tiputini Biodiversity Station (Amazonian Ecuador), 12
Tokuda, Marcos, 116–19, 202n11
tool use, and cultural transmission, 141–42, 144, 157–61, 205n4, 206n9
topi antelope (*Damaliscus lunatus*), 102–4, 200n11
travel networks, xii, 108–22, 184–85; betweenness in, 42; and grooming networks, 115; and leadership, 111–12, 121–22. *See also* migration
Tregenza, Tom, 79–81, 197nn6–7
Trivers, Robert, 44, 193n6
trophallaxis, 128–32, *gallery*

UNESCO (United Nations Educational, Scientific and Cultural Organization), 12, 113, 169

vampire bats (*Desmodus rotundus*), 43–49, 58, *gallery*
Verreaux's sifaka (*Propithecus verreauxi*), lemurs, 169–73, 182, 184, 206n4
vervet monkeys (*Chlorocebus pygerythrus*), 18, 23, 34–38, 192n14; grooming networks of, 35–36; juvenile, 36
von Frisch, Karl, 127, 130, 203n4

Wallace, Alfred Russel, 19
Wang, Qishan, 113
water dragons (*Intellagama lesueurii*), reptiles, 135, 204n11
Weidt, Andrea, 94–95
Whale Center of New England (Gloucester, MA), 146–48
whales, xii. *See also* humpback whales (*Megaptera novaeangliae*)
Wheelwright, Nathaniel Thoreau, 67
white storks. *See* European white storks (*Ciconia ciconia*)
Wikelski, Martin, 112
wildebeests, blue, 98, 103–4
Wilkinson, Gerald, 43–46, 193n6
Wingtags (Sydney, Australia), 24–28
wire-tailed manakins (*Pipra filicauda*), 11–15. *See also* long-tailed manakins (*Chiroxiphia linearis*)
Woodchester Park (Stroud, UK), 173–74, 177
Wytham Tit Project (Oxford), 152–54

Xia, Dong-Po, 113–15, 201–2n7

Yasuní Biosphere Reserve (Ecuador), 11–14